SpringerBriefs in Applied Sciences and Technology

SpringerBriefs present concise summaries of cutting-edge research and practical applications across a wide spectrum of fields. Featuring compact volumes of 50–125 pages, the series covers a range of content from professional to academic.

Typical publications can be:

- A timely report of state-of-the art methods
- An introduction to or a manual for the application of mathematical or computer techniques
- A bridge between new research results, as published in journal articles
- A snapshot of a hot or emerging topic
- An in-depth case study
- A presentation of core concepts that students must understand in order to make independent contributions

SpringerBriefs are characterized by fast, global electronic dissemination, standard publishing contracts, standardized manuscript preparation and formatting guidelines, and expedited production schedules.

On the one hand, **SpringerBriefs in Applied Sciences and Technology** are devoted to the publication of fundamentals and applications within the different classical engineering disciplines as well as in interdisciplinary fields that recently emerged between these areas. On the other hand, as the boundary separating fundamental research and applied technology is more and more dissolving, this series is particularly open to trans-disciplinary topics between fundamental science and engineering.

Indexed by EI-Compendex, SCOPUS and Springerlink.

More information about this series at http://www.springer.com/series/8884

Tin-Chih Toly Chen · Katsuhiro Honda

Fuzzy Collaborative Forecasting and Clustering

Methodology, System Architecture, and Applications

Tin-Chih Toly Chen
Department of Industrial Engineering
and Management
National Chiao Tung University
Hsinchu, Taiwan

Katsuhiro Honda
Graduate School of Engineering
Osaka Prefecture University
Sakai, Osaka, Japan

ISSN 2191-530X   ISSN 2191-5318   (electronic)
SpringerBriefs in Applied Sciences and Technology
ISBN 978-3-030-22573-5   ISBN 978-3-030-22574-2   (eBook)
https://doi.org/10.1007/978-3-030-22574-2

This Springer imprint is published by the registered company Springer Nature Switzerland AG
The registered company address is: Gewerbestrasse 11, 6330 Cham, Switzerland

Preface

Fuzzy systems have been successfully applied to various problems with uncertainty, including clustering, system control, decision making, and forecasting. However, most of these applications are based on a single fuzzy approach/system that is chosen in a subjective way. In addition, with the widespread of Internet applications, dealing with disparate data sources is becoming increasingly popular. Furthermore, due to technical limitations, security or privacy considerations, the integral access to a number of sources is often limited. For these reasons, the concepts of collaborative computing intelligence and collaborative fuzzy modeling have been proposed; the so-called fuzzy collaborative system have been developed. So far, several studies have argued that for certain problems, a fuzzy collaborative intelligence approach is more precise, accurate, efficient, safe, and private than typical approaches. Although there have been some literature about fuzzy collaborative intelligence and systems, considerable room for development still exists in this field. For example, a crisp method or system needs to be fuzzified to meet the requirements of a fuzzy collaborative intelligence method or system; the collaboration among the participating decision makers needs to be facilitated; and the views of, and the results by, the decision makers have to be aggregated.

So far, most existing fuzzy collaboration systems have been used for clustering, filtering, and forecasting. This book is dedicated to two interesting topics in fuzzy collaborative intelligence and systems, i.e., fuzzy collaborative forecasting and fuzzy collaborative clustering. Both fuzzy collaborative forecasting and fuzzy collaborative clustering are major types of fuzzy collaborative intelligence and systems. However, fuzzy collaborative forecasting is supervised learning because the actual value exists, while fuzzy collaborative clustering is unsupervised learning because there is no absolute clustering result.

It is necessary to acquire a general knowledge of the most useful fuzzy collaborative intelligence and systems in order to be able to apply them efficiently in real-life situations. To this end, six chapters have been provided in this book that belongs to the SpringerBriefs in Applied Sciences and Technology series.

Chapter 1 gives the definitions of fuzzy collaborative intelligence and fuzzy collaborative systems. Then, some existing fuzzy intelligence and systems are classified. The operation procedure of a fuzzy collaborative system is also detailed.

Chapter 2 introduces some linear fuzzy collaborative forecasting methods or models. Then, the steps in operating a fuzzy collaborative forecasting system, including collaboration, aggregation, and defuzzification, are detailed. How to assess the effectiveness of a fuzzy collaborative forecasting method and how to measure the quality of collaboration are also discussed.

Chapter 3 introduces several nonlinear fuzzy collaborative forecasting methods. In addition, a special application of nonlinear fuzzy collaborative forecasting, the collaborative fuzzy analytic hierarchy process, is also described in this chapter.

Chapter 4 reviews fuzzy c-Means (FCM) and its variants, which are the fundamental methods for unsupervised data classification. Then, fuzzy co-clustering, that are prevalent in cooccurrence information analysis, is introduced.

Chapter 5 reviews several collaborative clustering models and introduces a collaborative framework of fuzzy co-clustering. Two types of distributed databases, i.e., vertically distributed databases and horizontally distributed databases, can then be handled with different security concepts.

Chapter 6 reviews three-mode fuzzy co-clustering that reveals the intrinsic co-cluster structures from three-mode cooccurrence information. In addition, a framework for securely applying three-mode fuzzy co-clustering is also developed, when cooccurrence information is stored in different organizations.

The purpose of the book is not to be exhaustive in the list of methods and algorithms that exist in the relevant literature. It is intended to provide technical details of the development of fuzzy collaborative intelligence and systems and the corresponding applications. These details will hold great interest for researchers in information engineering, information management, artificial intelligence, and computational intelligence, as well as for practicing managers and engineers.

Hsinchu, Taiwan   Tin-Chih Toly Chen
Sakai, Osaka, Japan   Katsuhiro Honda

Contents

Chapter 1
Introduction to Fuzzy Collaborative Forecasting Systems

1.1 Fuzzy Collaborative Intelligence and Systems

Multiple analyses of a problem from diverse perspectives raise the chance that no relevant aspects of the problem will be ignored. In addition, as Internet applications become widespread, dealing with disparate data sources is becoming more and more popular. Technical constraints, security issues, and privacy considerations often limit access to some sources. Therefore, the concepts of collaborative computing intelligence and collaborative fuzzy modeling have been proposed, and certain so-called fuzzy collaborative systems are being established [1, 2]. In a fuzzy collaborative system, some experts, agents, or systems with various backgrounds are trying to achieve a common target. Since they have different knowledge and points of view, they may use various methods to model, identify, or control the common target. The key to such a system is that these experts, agents, or systems share and exchange their observations, settings, experiences, and knowledge each other when achieving the common goal. This features the fuzzy collaborative system distinct from the ensemble of multiple fuzzy systems.

Although there have been some literature about fuzzy collaborative intelligence and systems, considerable room for development still exists in this field, for example,

- how to fuzzify a method or system to meet the requirements of a fuzzy collaborative intelligence method or system,
- how to facilitate the collaboration among experts, and
- how to aggregate the views of, and the results by, experts.

So far, most existing fuzzy collaboration systems have been used for clustering [3, 4], filtering [5, 6], and forecasting [7, 8], as illustrated in Fig. 1.1. Some of the relevant literature in these fields are briefly reviewed as follows.

© The Author(s), under exclusive license to Springer Nature Switzerland AG 2020
T.-C. T. Chen and K. Honda, *Fuzzy Collaborative Forecasting and Clustering*,
SpringerBriefs in Applied Sciences and Technology,
https://doi.org/10.1007/978-3-030-22574-2_1

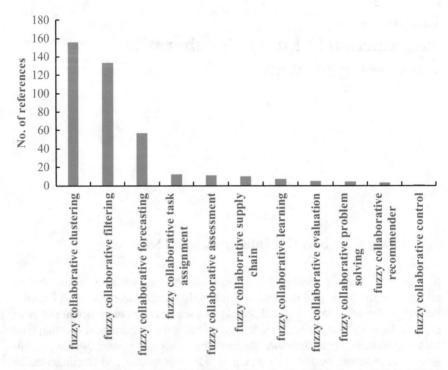

Fig. 1.1 Number of references about fuzzy collaborative intelligence and systems from 2000 to 2018. *Data source* Google Scholar

Clustering is the task of grouping (clustering) a set of objects in such a way that objects in the same cluster are more similar to each other than to those in other clusters. In fuzzy clustering, an object can belong to multiple clusters to different degrees. Clustering itself is a task involving much uncertainty, since there is no absolute rule for forming clusters. Once an object can be assigned to multiple clusters, the clustering results will be more diverse, which requires collaboration and aggregation into a single, representative result. Pedrycz and Rai [3] discussed a collaborative data clustering problem in which data were scattered at multiple data sites. There was an agent on each data site that clustered the data on the data site. As a result, each agent had access to only a part of the data and constructed a partitioning matrix to cluster the data part. Subsequently, agents communicated their partitioning matrices and structure findings to each other, so that each agent could modify the clustering result by incorporating other agents' partitioning matrices and structure findings into the objective function of the optimization problem for forming clusters. In the view of Mitra et al. [4], fuzzy collaborative clustering is an implementation of the divide-and-conquer strategy that is often adopted to enable efficient mining of large databases. In addition, the number of clusters on a data site could be different before and after collaboration.

Clustering Web pages facilitates the recommendation of related Web pages to users. Web pages can be clustered according to their attributes. In addition, Web pages may correlate with each other, which provide useful information on the clustering of Web pages. From this point of view, Loia et al. [5] proposed an extension of the well-known fuzzy c-Means (FCM) algorithm, namely the proximity FCM (P-FCM) algorithm, to cluster Web pages. The P-FCM algorithm ran FCM iteratively. In each iteration, the similarity or dissimilarity between two Web pages from a user's feedback was incorporated to adjust the clustering result. The iteration process continued until the distance between two successive clustering results had been negligible.

Collaborative filtering is a recommendation technique that makes a recommendation based on the similarity between the cases. In a fuzzy collaborative filtering method, the similarity between two cases is modeled with a fuzzy number [6]. In addition, similarity can be measured along various dimensions. As a result, the measurement results by considering different dimensions may not be the same and need to be aggregated using techniques such as the weighted sum [6]. Finally, for a new case, the recommendation made to the old case that is most similar to the new case is recommended. Another way is to consider multiple dimensions simultaneously when measuring similarity. For example, Leung et al. [7] mined fuzzy association rules from the attributes/dimensions and choices of old cases. Then, for a new case, the fuzzy association rules fired were applied to make recommendations. The memberships of the same alternative made by all fuzzy association rules were added up. Finally, the alternative with the highest sum of memberships is recommended.

Forecasting is the process of making predictions of the future based on past and present data, usually by analyzing the trends. In fuzzy forecasting, fuzzy methods are applied to generate forecasts that may be expressed in crisp or fuzzy values. Prevalent fuzzy inference systems (FISs) are an example of fuzzy collaborative forecasting systems [8]. In a Mamdani FIS, several fuzzy inference rules forecast the same target from different points of view. The forecasts by the fuzzy inference rules are aggregated using the union operator. Then, the aggregation result is defuzzified using the center of gravity method. However, it is not guaranteed that the aggregation result contains the actual value. To address this issue, Chen [9] proposed a hybrid fuzzy linear regression (FLR)-back-propagation network (BPN) approach to predict the efficient cost per unit of a semiconductor product. In the FLR-BPN approach, a group of experts was formed. Each expert fitted a FLR equation to predict the unit cost by solving a mathematical programming problem, so that the fuzzy forecast by each expert contained the actual value. All fuzzy forecasts were aggregated using fuzzy intersection, resulting in a polygon-shaped fuzzy number that was defuzzified using a BPN.

Chen [10] considered the case in which each expert has only partial access to the data, and is not willing to share the raw data he/she owns. The forecasting results by an expert are conveyed to other experts for the modification of their settings, so that the actual values will be contained in the fuzzy forecasts after collaboration.

So far, several studies have argued that for certain problems, a fuzzy collaborative intelligence approach is more precise, accurate, efficient, safe, and private than typical approaches. However, there are still some issues that need to be addressed. According to Poler et al. [11], the comparison of collaboration methods and the proposing of software tools, especially as regards forecasting methods for collaborative forecasting, are still lacking. All fuzzy collaborative intelligence methods seek the consensus of results. In this field, Ostrosi et al. [12] defined the concept of consensus as the overlapping of design clusters of different perspectives. Similarly, Chen [13] defined the concept of partial consensus as the intersection of the views of some experts. A different aggregation mechanism was proposed by [14, 15] that minimized the sum of the squared deviations between each expert's judgement and the aggregation result. However, the aggregation mechanism assumed the existence of consensus and derived the aggregation result directly. Cheikhrouhou et al. [16] thought that collaboration is necessary because of the unexpected events that may occur in the future demand.

This book is dedicated to two interesting topics in fuzzy collaborative intelligence and systems, i.e., fuzzy collaborative forecasting and fuzzy collaborative clustering. Both fuzzy collaborative forecasting and fuzzy collaborative clustering are major types of fuzzy collaborative intelligence and systems. However, fuzzy collaborative forecasting is supervised learning because the actual value exists, while fuzzy collaborative clustering is unsupervised learning because there is no absolute clustering result.

This book aims to introduce the basic concepts of fuzzy collaborative forecasting and fuzzy collaborative clustering, including methodology, system architecture, and applications. In specific, the outline of the present book is structured as follows: In the current chapter, first, the definitions of fuzzy collaborative intelligence and fuzzy collaborative system are given. Then, existing fuzzy intelligence and systems are classified. The operation procedure of a fuzzy collaborative system is also detailed. Chapter 2, Linear Fuzzy Collaborative Forecasting Methods, starts from the review of some linear fuzzy forecasting methods. Based on these linear fuzzy forecasting methods, some linear fuzzy collaborative forecasting methods or models are introduced. Subsequently, the steps in operating a fuzzy collaborative forecasting system, including collaboration, aggregation, and defuzzification, are detailed. To assess the effectiveness of applying a fuzzy collaborative forecasting method, how to measure the quality of collaboration is also discussed. In Chap. 3, Nonlinear Fuzzy Collaborative Forecasting Methods, three nonlinear fuzzy forecasting methods, i.e., the modified back-propagation network approach, the fuzzy back-propagation network approach, and the simplified calculation technique for the fuzzy back-propagation network approach, are first introduced. Nonlinear fuzzy collaborative forecasting methods are proposed based on these nonlinear fuzzy forecasting methods. In addition, a special application of nonlinear fuzzy collaborative forecasting, the collaborative fuzzy analytic hierarchy process, is also mentioned in this chapter, in which the priorities (or weights) of factors (attributes, or criteria) that are acceptable to all decision makers are to be estimated. The collaborative fuzzy analytic hierarchy process is considered as an unsupervised fuzzy collaborative forecasting problem, since there are no actual values of the fuzzy priorities. Chapter 4, Fuzzy Clustering and Fuzzy Co-clustering, gives a brief review on the basic fuzzy clustering model of fuzzy c-Means (FCM) and its variants, which are the fundamental methods for unsupervised data classification. And then, fuzzy co-clustering is introduced, which are available in cooccurrence information analysis such as document–keyword frequencies and customer-product purchase history transactions. The FCM-type objective function is modified with the cluster-wise aggregation degrees among objects and items. In Chap. 5, Collaborative Framework for Fuzzy Co-clustering, several collaborative clustering models are reviewed followed by an introduction of a collaborative framework of fuzzy co-clustering. In collaborative clustering, two types of distributed databases, i.e., vertically distributed databases and horizontally distributed databases, are handled with different security concepts. Finally, Chap. 6, Three-mode Fuzzy Co-clustering and Collaborative Framework, conducts a brief review on three-mode fuzzy co-clustering, which reveals the intrinsic co-cluster structures from three-mode cooccurrence information. Additionally, how to develop a framework of securely applying three-mode fuzzy co-clustering is described, where cooccurrence information is stored in different organizations.

1.2 Classification of Fuzzy Collaborative Intelligence and Systems

Fuzzy collaborative intelligence and systems can be classified based on three dimensions [17]:

(1) The decentralized or centralized data access,
(2) The use of a real expert or software agent (virtual expert): Domain experts must be asked to give their opinions in a fuzzy collaborative intelligence approach, so that the results generated can satisfy the requirements of these experts. Nevertheless, in case that there are no domain experts to consult, there is always the alternative of forming a committee of "virtual experts" for providing opinion sets for the fuzzy collaborative intelligence approach. This does not mean that opinion sets can be randomly generated, because impractical opinion sets might result in no feasible solutions of the optimization problems or increase the difficulty to find an optimal solution.
(3) The similarities and differences between the fuzzy methods (or systems) used on different locations (or by different experts) (refer to Fig. 1.2).

In particular, when the fuzzy methods applied by experts are of the same type, the fuzzy collaborative intelligence method is called a homogeneous fuzzy collaborative intelligence method; otherwise, the fuzzy collaborative intelligence method is called a heterogeneous fuzzy collaborative intelligence method [18]. Most existing methods are homogeneous, although inspecting a problem from various perspectives ensures that no major component of the problem is ignored. On the basis of this belief, a heterogeneous method that approaches the problem from various perspectives would be valuable. For example, in Chen [18], a heterogeneous fuzzy collaborative intelligence approach was proposed for forecasting the future yield of a product. In the

Fig. 1.2 Dimensions of
fuzzy collaborative
intelligence and systems

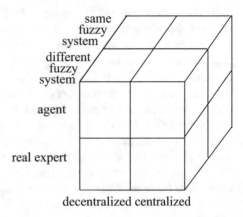

decentralized centralized

approach, in addition to the use of prevalent mathematical programming methods, an artificial neural network was also constructed to fit a yield learning process.

1.3   Operating Procedure of a Fuzzy Collaborative System

The operating procedure of a fuzzy collaborative system starts from the formation of the expert (or software agent) group and usually consists of the following steps [19, 20] (see Fig. 1.3):

1. Each expert (or software agent) determines the setting of the fuzzy method he/she will apply: The fuzzy methods applied by different experts may not be the same (or of the same type). Fuzzy methods are homogeneous if they are of the same type. For example, in Chen and Lin [21], the fuzzy forecasting methods applied by all experts are based on mathematical programming models. Fuzzy methods are heterogeneous if they are of different types. For example, in Chen [18], two types of fuzzy forecasting methods, one based on mathematical programming and the other based on artificial neural network, were applied. When agents are used, the setting will be casual and less practical.
2. Each expert applies the fuzzy method to fulfill the task.
3. Each expert communicates his/her view and results to other experts, perhaps with the aid of a centralized collaboration mechanism.
4. Upon receipt, the expert adjusts his/her setting and re-applies the fuzzy method to fulfill the task.
5. The results by all experts are aggregated.
6. The aggregation result is defuzzified to generate a single, representative value.
7. The overall performance is evaluated.
8. The collaboration process is terminated if the improvement in the overall performance becomes negligible. Otherwise, return to Step 3.

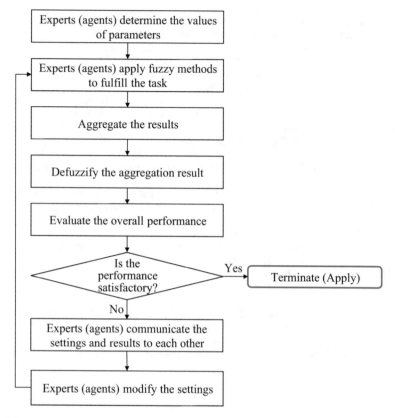

Fig. 1.3 Operating procedure of a fuzzy collaborative system

References

1. W. Pedrycz, Collaborative fuzzy clustering. Pattern Recognit. Lett. **23**, 1675–1686 (2002)
2. W. Pedrycz, Collaborative architectures of fuzzy modeling. Lect. Notes Comput. Sci. **5050**, 117–139 (2008)
3. W. Pedrycz, P. Rai, A multifaceted perspective at data analysis: a study in collaborative intelligent agents. IEEE Trans. Syst. Man Cybern. Part B (Cybernetics) **38**(4), 1062–1072 (2008)
4. S. Mitra, H. Banka, W. Pedrycz, Rough–fuzzy collaborative clustering. IEEE Trans. Syst. Man Cybern. Part B (Cybernetics) **36**(4), 795–805 (2006)
5. V. Loia, W. Pedrycz, S. Senatore, P-FCM: a proximity-based fuzzy clustering for user-centered web applications. Int. J. Approx. Reason. **34**, 121–144 (2003)
6. L.H. Son, HU-FCF: a hybrid user-based fuzzy collaborative filtering method in recommender systems. Expert Syst. Appl. Int. J. **41**(15), 6861–6870 (2014)
7. C.W.K. Leung, S.C.F. Chan, F.L. Chung, A collaborative filtering framework based on fuzzy association rules and multiple-level similarity. Knowl. Inf. Syst. **10**(3), 357–381 (2006)
8. A. Amindoust, S. Ahmed, A. Saghafinia, A. Bahreininejad, Sustainable supplier selection: a ranking model based on fuzzy inference system. Appl. Soft Comput. **12**(6), 1668–1677 (2012)
9. T. Chen, An effective fuzzy collaborative forecasting approach for predicting the job cycle time in wafer fabrication. Comput. Ind. Eng. **66**(4), 834–848 (2013)

10. T. Chen, An application of fuzzy collaborative intelligence to unit cost forecasting with partial data access by security consideration. Int. J. Technol. Intell. Plann. **7**(3), 201–214 (2011)
11. R. Poler, J.E. Hernandez, J. Mula, F.C. Lario, Collaborative forecasting in networked manufacturing enterprises. J. Manuf. Technol. Manage. **19**(4), 514–528 (2008)
12. E. Ostrosi, L. Haxhiaj, S. Fukuda, Fuzzy modelling of consensus during design conflict resolution. Res. Eng. Design **23**(1), 53–70 (2012)
13. T. Chen, A hybrid fuzzy and neural approach with virtual experts and partial consensus for DRAM price forecasting. Int. J. Innov. Comput. Inf. Control **8**(1), 583–597 (2012)
14. E. Ostrosi, J.B. Bluntzer, Z. Zhang, J. Stjepandić, Car style-holon recognition in computer-aided design. J. Comput. Des. Eng. article in press (2018)
15. Z. Zhang, D. Xu, E. Ostrosi, L. Yu, B. Fan, A systematic decision-making method for evaluating design alternatives of product service system based on variable precision rough set. J. Intell. Manuf. article in press (2017)
16. N. Cheikhrouhou, F. Marmier, O. Ayadi, P. Wieser, A collaborative demand forecasting process with event-based fuzzy judgements. Comput. Ind. Eng. **61**(2), 409–421 (2011)
17. T. Chen, A collaborative fuzzy-neural system for global CO_2 concentration forecasting. Int. J. Innov. Comput. Inf. Control **8**(11), 7679–7696 (2012)
18. T. Chen, A heterogeneous fuzzy collaborative intelligence approach for forecasting the product yield. Appl. Soft Comput. **57**, 210–224 (2017)
19. T. Chen, Y.C. Lin, A fuzzy-neural system incorporating unequally important expert opinions for semiconductor yield forecasting. Int. J. Uncertainty Fuzziness Knowledge-Based Syst. **16**(01), 35–58 (2008)
20. T. Chen, An agent-based fuzzy collaborative intelligence approach for predicting the price of a dynamic random access memory (DRAM) product. Algorithms **5**(2), 304–317 (2012)
21. T. Chen, Y.C. Wang, An agent-based fuzzy collaborative intelligence approach for precise and accurate semiconductor yield forecasting. IEEE Trans. Fuzzy Syst. **22**(1), 201–211 (2014)

Chapter 2
Linear Fuzzy Collaborative Forecasting Methods

2.1 Linear Fuzzy Forecasting Methods

Linear methods have been widely applied to forecasting. Prevalent linear forecasting methods include moving average, exponential smoothing, linear regression (LR), autoregressive integrated moving average (ARIMA), and others. Fuzzifying the parameters of a linear forecasting method changes it to a linear fuzzy forecasting method.

Various linear fuzzy forecasting methods have been proposed in the past. Among them, fuzzy linear regression (FLR) is one of the most commonly applied methods. For example, Song et al. [1] constructed a FLR equation to forecast the short-term load of a holiday, which was a typical time series problem. Inputs to the FLR equation were the load data of the previous three weekdays, including the average values and standard deviations that were approximated with symmetric triangular fuzzy numbers (TFNs). The output was the load forecast of the holiday also expressed with a symmetric TFN. The coefficients in the FLR equation were derived by solving a linear programming (LP) problem. A symmetric fuzzy forecast was comparable to a confidence interval in statistics, which made Song et al.'s method attractive. However, the product (multiplication) of symmetric TFNs was no longer symmetric, which meant some simplification was done in the LP model. FLR or possibilistic linear regression methods have also been extensively applied to model learning processes that involve uncertainty [2]. For example, in Chen and Wang [3], after converting all parameters to their logarithmic values, an uncertain yield learning process could be fitted with a FLR equation. The fuzzy parameters in the FLR equation were asymmetric. A LP model, different from that proposed by Song et al. [1], was also formulated to derive the values of the fuzzy parameters. Tseng et al. [4] proposed a

© The Author(s), under exclusive license to Springer Nature Switzerland AG 2020

T.-C. T. Chen and K. Honda, *Fuzzy Collaborative Forecasting and Clustering*,
SpringerBriefs in Applied Sciences and Technology,
https://doi.org/10.1007/978-3-030-22574-2_2

Fig. 2.1 Two extremes of
fuzzy forecasting objectives

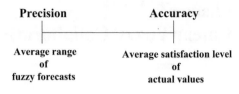

Fig. 2.2 Architecture of a
linear fuzzy forecasting
method

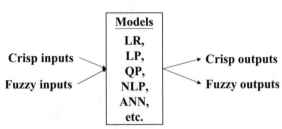

fuzzy ARIMA (FARIMA) approach for forecasting the foreign exchange rate. They first derived the values of parameters in the ARIMA model with autocorrelation function and partial autocorrelation function and then solved a LP problem to fuzzify the ARIMA model.

The aim of a linear fuzzy forecasting method is to improve both the accuracy and precision of forecasting the target [5]:

(1) Accuracy: The forecasted value should be as close as possible to the actual value.
(2) Precision: A narrow interval containing the actual value is established.

To these ends, various approaches, including different objective functions and/or model formulations, have been adopted to fuzzify a linear forecasting method, i.e., to derive the values of parameters in the linear fuzzy forecasting method. In these approaches, some aim to optimize the forecasting precision by minimizing the average range of fuzzy forecasts [3, 6], while others optimize the forecasting accuracy by maximizing the average satisfaction level (i.e., the average membership of actual values in the corresponding fuzzy forecasts) [7], as illustrated in Fig. 2.1.

Inputs to a linear fuzzy forecasting method can be crisp or fuzzy values. Subsequently, the values of fuzzy parameters can be derived by solving a LR [4], LP [3, 6, 7], quadratic programming (QP) [8], nonlinear programming (NLP) [9], or artificial neural network (ANN) training [10] problem. Finally, the outputs from the linear fuzzy forecasting method can also be crisp or fuzzy values, as illustrated in Fig. 2.2. Among the possible types, a linear fuzzy forecasting method with crisp inputs and fuzzy outputs is the most commonly applied type. The FLR method, introduced in the next section, is an example of this type of linear fuzzy forecasting methods.

2.2 The Fuzzy Linear Regression (FLR) Method

A FLR equation with crisp inputs and a fuzzy output has the following form:

$$\tilde{y}_i = \tilde{w}_0(+) \sum_{k=1}^{K} \tilde{w}_k x_{ik} \tag{2.1}$$

where x_{ik} is the k-th input of the i-th example; $k = 1 \sim K$; $i = 1 \sim n$. $\tilde{y}_i$ is the output (i.e., the fuzzy forecast). $\tilde{w}_k$ is the coefficient for x_{ik}. (+) indicates fuzzy multiplication. When all fuzzy parameters are given in TFNs, the output is also a TFN:

$$\tilde{y}_i = (y_{i1}, y_{i2}, y_{i3}) \tag{2.2}$$

where

$$y_{i1} = w_{01} + \sum_{k=1}^{K} w_{k1} x_{ik} \tag{2.3}$$

$$y_{i2} = w_{02} + \sum_{k=1}^{K} w_{k2} x_{ik} \tag{2.4}$$

$$y_{i3} = w_{03} + \sum_{k=1}^{K} w_{k3} x_{ik} \tag{2.5}$$

The following LP model derives the values of fuzzy parameters by minimizing the average range of fuzzy forecasts [3, 6]:
(LP Model I)

$$\text{Min } Z_1 = \sum_{i=1}^{n} (y_{i3} - y_{i1}) \tag{2.6}$$

subject to

$$a_i \geq (1 - s)y_{i1} + s y_{i2}; \; i = 1 \sim n \tag{2.7}$$

$$a_i \leq (1 - s)y_{i3} + s y_{i2}; \; i = 1 \sim n \tag{2.8}$$

$$y_{i1} = w_{01} + \sum_{k=1}^{K} w_{k1} x_{ik}; \; i = 1 \sim n \tag{2.9}$$

$$y_{i2} = w_{02} + \sum_{k=1}^{K} w_{k2} x_{ik}; \; i = 1 \sim n \tag{2.10}$$

Fig. 2.3 Satisfaction level

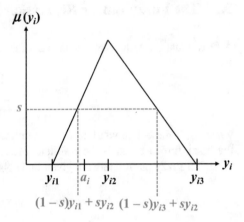

$$(1-s)y_{i1} + sy_{i2} \quad (1-s)y_{i3} + sy_{i2}$$

$$y_{i3} = w_{03} + \sum_{k=1}^{K} w_{k3}x_{ik}; \quad i = 1 \sim n \tag{2.11}$$

$$w_{k1} \leq w_{k2} \leq w_{k3}; \quad k = 0 \sim K \tag{2.12}$$

where a_i is the actual value; s is the pre-specified satisfaction level within $[0,1]$. Constraints (2.7) and (2.8) ensure that the membership of the actual value in the fuzzy forecast is greater than s, as illustrated in Fig. 2.3. A larger value of s results in a wider fuzzy forecast. A fuzzy forecast contains the actual value, which is the basis for the subsequent collaboration.

By maximizing the average satisfaction level instead, the following QP model can also be used to derive the values of fuzzy parameters [7]:

(QP Model I)

$$\text{Max } Z_2 = \sum_{i=1}^{n} s_i \tag{2.13}$$

subject to

$$\frac{1}{n} \sum_{i=1}^{n} (y_{i3} - y_{i1}) \leq d; \quad i = 1 \sim n \tag{2.14}$$

$$a_i \geq (1 - s_i)y_{i1} + s_i y_{i2}; \quad i = 1 \sim n \tag{2.15}$$

$$a_i \leq (1 - s_i)y_{i3} + s_i y_{i2}; \quad i = 1 \sim n \tag{2.16}$$

$$y_{i1} = w_{01} + \sum_{k=1}^{K} w_{k1}x_{ik}; \quad i = 1 \sim n \tag{2.17}$$

$$y_{i2} = w_{02} + \sum_{k=1}^{K} w_{k2} x_{ik}; \quad i = 1 \sim n \tag{2.18}$$

$$y_{i3} = w_{03} + \sum_{k=1}^{K} w_{k3} x_{ik}; \quad i = 1 \sim n \tag{2.19}$$

$$w_{k1} \leq w_{k2} \leq w_{k3}; \quad k = 0 \sim K \tag{2.20}$$

$$0 \leq s_i \leq 1; \quad i = 1 \sim n \tag{2.21}$$

Another QP model for the same purpose was proposed by Donoso et al. [8]:
(**QP Model II**)

$$\text{Min } Z_3 = \omega_1 \sum_{i=1}^{n} (y_{i2} - a_i)^2 + \omega_2 \sum_{i=1}^{n} (y_{i3} - y_{i1} - \Delta y_i)^2 \tag{2.22}$$

subject to the same constraints with LP Model I. The objective function is to minimize the weighted sum of the squared deviations from the cores (i.e., y_{i2}) and the squared deviations from the estimated spreads (i.e., Δy_i). ω_1 and ω_2 are the weights assigned to the two terms. $\omega_1, \omega_2 \in [0, 1]$; $\omega_1 + \omega_2 = 1$.

Several methods can be applied to solve a QP problem, such as the interior point method, the active set method, the augmented Lagrangian method, the conjugate gradient method, the gradient projection method, and the extensions of the simplex algorithm [11]. However, it is not always easy to find the global optimal solutions to the QP models. A fuzzy collaborative forecasting method evolves a fuzzy forecast from those made by multiple experts. Whether the fuzzy forecasts made by the experts are good or bad, locally or globally optimal, is not very important. The only prerequisite for the fuzzy forecasts is the inclusion of the actual value (for the training data). Obviously, all feasible solutions to the QP models meet this requirement. In contrast, it will be better if the fuzzy forecasts made by the experts are diversified.

The fuzzy forecasts generated using different models (or by different experts) are not the same and, therefore, can be aggregated to arrive at a better fuzzy forecast.

2.3 The Operational Procedure of a Fuzzy Collaborative Forecasting System

The operational procedure of a fuzzy collaborative forecasting system consists of several steps that will be described in the following sections (see Fig. 2.4):

- Step 1. The fuzzy collaborative forecasting system starts from the formation of a group of domain experts (or software agents).

- Step 2. The experts put forward their views on certain aspects of forecasting that are incorporated into the fuzzy forecasting methods they will apply.
- Step 3. Each expert forecasts the target using the fuzzy forecasting method based on his/her view.
- Step 4. Each expert conveys his/her view and forecasting results to others with the aid of a centralized control unit based on the collaboration mechanisms.
- Step 5. The centralized control unit aggregates the fuzzy forecasts by all experts to arrive at a representative value.
- Step 6. The centralized control unit defuzzifies the aggregation result, so as to assess the overall forecasting performance.
- Step 7. After receiving the views and forecasting results of others, a domain expert may be affected to modify his/her view.
- Step 8. The collaboration process is terminated if the improvement in the overall forecasting performance becomes negligible. Otherwise, return to Step (3).

2.4   A Linear Fuzzy Collaborative Forecasting Method

Linear fuzzy collaborative forecasting methods are fuzzy collaborative forecasting methods in which participants (either real experts or software agents) apply linear fuzzy forecasting methods. If participants apply different linear fuzzy forecasting methods, or the same linear fuzzy forecasting method but with different settings, then the fuzzy forecasts will not be equal [9]. For example, the following parameters can be set for the models introduced above:

- The sensitivity to uncertainty (o): In LP Model I, by minimizing the sum of the ranges/supports of fuzzy forecasts, it is not very sensitive to uncertainty. Minimizing a higher-order sum of ranges instead theoretically elevates the sensitivity to uncertainty and might reflect the attitudes of experts better. Conversely, it is also possible to minimize a lower-order sum instead.
- The desired range of a fuzzy forecast (d): In QP Model I, the average range of fuzzy forecasts is restricted, so as to control uncertainty. The desired ranges by various experts might not be equal, which should be taken into account.
- The required satisfaction level (s): In LP Model I and QP Model II, the historical data are fitted at or above a given level of satisfaction. The required satisfaction levels by various experts are different in nature, which needs to be reflected in the model.
- The relative importance of the outliers of the sample data (m): In both models, the range of a fuzzy forecast is determined by outliers that are satisfied with the lowest levels (see Fig. 2.5). One way to tackle this problem is to find out and then to remove outliers [12]. Another way is to lessen the relative importance of outliers. For example, in QP Model II, the average satisfaction level ($\bar{s}$) can be calculated using the generalized mean instead:

Fig. 2.4 Operational
procedure of a fuzzy
collaborative forecasting
system

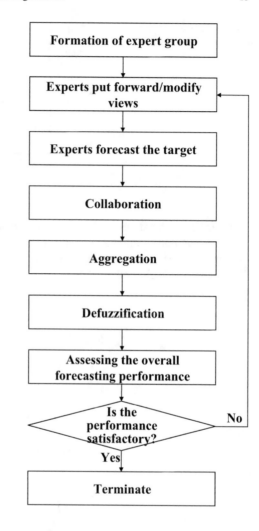

$$\bar{s} = \sqrt[m]{\frac{\sum_{i=1}^{n} s_i^m}{n}} \qquad (2.23)$$

where $m \in \mathbf{R}^+$; $m > 1$. For example, assume there are two data points satisfied
with different levels, 0.2 and 0.6, respectively. According to the original formula
($m = 1$), $\bar{s} = 0.4$. If the new formula (with $m = 2$) is used instead, then $\bar{s} = 0.45$.
The new result is farther from that of the outlier, which indicates that the relative
importance of the outlier is lessened after applying the generalized mean.

- Different priorities (ω_1 and ω_2): In QP Model II, the priorities (or weights) for
the squared deviations from the cores and for the squared deviations from the
estimated spreads specified by experts may be different.

Fig. 2.5 Ranges of fuzzy forecasts determined by outliers

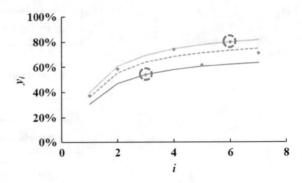

After taking the aforementioned issues into account, a linear FCF method incorporating unequally important expert views (or settings) is constructed. The linear FCF method incorporates expert views on the four issues and is composed of the following three NLP models:

(NLP Model I)

$$\text{Min } Z_4 = \sum_{i=1}^{n} (y_{i3} - y_{i1})^o \tag{2.24}$$

subject to

$$a_i \geq (1 - s)y_{i1} + sy_{i2}; \quad i = 1 \sim n \tag{2.25}$$

$$a_i \leq (1 - s)y_{i3} + sy_{i2}; \quad i = 1 \sim n \tag{2.26}$$

$$y_{i1} = w_{01} + \sum_{k=1}^{K} w_{k1}x_{ik}; \quad i = 1 \sim n \tag{2.27}$$

$$y_{i2} = w_{02} + \sum_{k=1}^{K} w_{k2}x_{ik}; \quad i = 1 \sim n \tag{2.28}$$

$$y_{i3} = w_{03} + \sum_{k=1}^{K} w_{k3}x_{ik}; \quad i = 1 \sim n \tag{2.29}$$

$$w_{k1} \leq w_{k2} \leq w_{k3}; \quad k = 0 \sim K \tag{2.30}$$

The order of the objective function is determined by o. If this is a large value, it increases the difficulty in optimizing the NLP problem. For this reason, o can be restricted to be less than 4. The range $[0, 4]$ allows for enough flexibility to formulate the NLP model.

(NLP Model II)

$$\text{Max } Z_5 = \bar{s} \tag{2.31}$$

subject to

$$\bar{s} = \sqrt[m]{\frac{\sum_{i=1}^{n} s_i^m}{n}} \tag{2.32}$$

$$\frac{1}{n} \sum_{i=1}^{n} (y_{i3} - y_{i1})^o \leq d^o; \quad i = 1 \sim n \tag{2.33}$$

$$a_i \geq (1 - s_i)y_{i1} + s_i y_{i2}; \quad i = 1 \sim n \tag{2.34}$$

$$a_i \leq (1 - s_i)y_{i3} + s_i y_{i2}; \quad i = 1 \sim n \tag{2.35}$$

$$y_{i1} = w_{01} + \sum_{k=1}^{K} w_{k1} x_{ik}; \quad i = 1 \sim n \tag{2.36}$$

$$y_{i2} = w_{02} + \sum_{k=1}^{K} w_{k2} x_{ik}; \quad i = 1 \sim n \tag{2.37}$$

$$y_{i3} = w_{03} + \sum_{k=1}^{K} w_{k3} x_{ik}; \quad i = 1 \sim n \tag{2.38}$$

$$w_{k1} \leq w_{k2} \leq w_{k3}; \quad k = 0 \sim K \tag{2.39}$$

$$0 \leq s_i \leq 1; \quad i = 1 \sim n \tag{2.40}$$

(NLP Model III)

$$\text{Min } Z_6 = \omega_1 \sum_{i=1}^{n} (y_{i2} - a_i)^o + \omega_2 \sum_{i=1}^{n} (y_{i3} - y_{i1} - \Delta y_i)^o \tag{2.41}$$

subject to

$$a_i \geq (1 - s)y_{i1} + s y_{i2}; \quad i = 1 \sim n \tag{2.42}$$

$$a_i \leq (1 - s)y_{i3} + s y_{i2}; \quad i = 1 \sim n \tag{2.43}$$

$$y_{i1} = w_{01} + \sum_{k=1}^{K} w_{k1} x_{ik}; \quad i = 1 \sim n \tag{2.44}$$

$$y_{i2} = w_{02} + \sum_{k=1}^{K} w_{k2}x_{ik}; \quad i = 1 \sim n \tag{2.45}$$

$$y_{i3} = w_{03} + \sum_{k=1}^{K} w_{k3}x_{ik}; \quad i = 1 \sim n \tag{2.46}$$

$$w_{k1} \leq w_{k2} \leq w_{k3}; \quad k = 0 \sim K \tag{2.47}$$

The values of these parameters chosen by the g-th expert, i.e., the parametric setting by the expert, are indicated with $o_{(g)}, s_{(g)}, d_{(g)}, m_{(g)}, \omega_{(g)1}$ and $\omega_{(g)2}$, respectively. $g = 1 \sim G$. The fuzzy forecast made by the expert is denoted by $\tilde{y}_i(g)$.

2.5 Collaboration Mechanisms

Reaching consensus is a crucial objective in multicriteria and multiexpert decision making [13]. The target of a collaboration mechanism is to help experts to reach a consensus. There are two perspectives for this: geometric representation and game theoretical modeling [13].

The consensus enhancing process is managed by an external supervisor (or a centralized control mechanism). In the beginning, each expert (or agent) can freely decide the adjustable parameters, which are then fed into the models to be solved. The expert then communicates the setting and forecasting results to others, perhaps with the aid of a centralized control unit. After receiving that information, the expert decides/modifies the adjustable parameters based on the information received, in order to improve the overall forecasting performance. To this end, Chen [14] proposed two rules:

- The along-the-favorable-direction rule: Assume the original views of two experts are denoted with $(o_{(1)}, s_{(1)}, d_{(1)}, m_{(1)}, \omega_{(1)1}, \omega_{(1)2})$ and $(o_{(2)}, s_{(2)}, d_{(2)}, m_{(2)}, \omega_{(2)1}, \omega_{(2)2})$, respectively. Expert #1 is favorable to the view of expert #2. Then according to the along-the-favorable-direction rule, the view of expert #1 will be modified in the following way:

$$(o_{(1)}, s_{(1)}, d_{(1)}, m_{(1)}, \omega_{(1)1}, \omega_{(1)2}) \rightarrow (o_{(1)} + \eta(o_{(2)} - o_{(1)}), s_{(1)} + \eta(s_{(2)} - s_{(1)}),$$
$$d_{(1)} + \eta(d_{(2)} - d_{(1)}), m_{(1)} + \eta(m_{(2)} - m_{(1)}), \omega_{(1)1} + \eta(\omega_{(2)1} - \omega_{(1)1}),$$
$$\omega_{(1)2} + \eta(\omega_{(2)2} - \omega_{(1)2}) \tag{2.48}$$

  where $0 \leq \eta \leq 1$.
- The proportional-to-the-original-value rule: According to the proportional-to-the-original-value rule, the view of expert #1 will be modified in the following way:

$$(o_{(1)}, s_{(1)}, d_{(1)}, m_{(1)}, \omega_{(1)1}, \omega_{(1)2}) \rightarrow (o_{(1)}(1 + \eta), s_{(1)}(1 + \eta), d_{(1)}(1 + \eta),$$

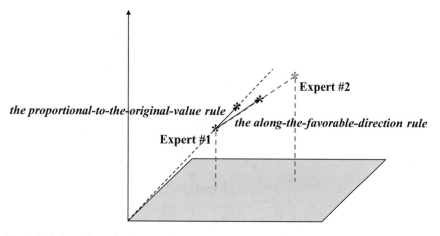

Fig. 2.6 A three-dimensional example for the two rules

$$m_{(1)}(1 + \eta), \omega_{(1)1}(1 + \eta), \omega_{(1)2}(1 + \eta)) \tag{2.49}$$

where $\eta \in \mathbf{R}$. The value of η is chosen so that the view moves closer to the favored one.

A three-dimensional example is provided in Fig. 2.6 to illustrate the two rules. If the fuzzy forecasts by experts converge, i.e., remain unchanged after some epochs, then the collaboration process should be stopped. On the other hand, if the improvement in the overall forecasting performance becomes negligible, then it is also unnecessary to continue the collaboration process.

In addition, the authorities of experts in forecasting the target can be differentiated. For example, in Chen [14], an expert's authority was determined with the following procedure:

- In the beginning, experts are of equal authorities. Namely, the authority of each expert is set to 1.
- Summarize the forecasting results by all experts and present them to every expert.
- If an expert favors the forecasting results of another expert, then add 0.5 to the authority of the latter.

which is illustrated in Fig. 2.7. Fuzzy forecasts made by experts of higher authorities will be emphasized more when aggregating the fuzzy forecasts.

2.6 Aggregation Mechanism

The fuzzy forecasts made by different experts are not equal, which requires aggregation and collaboration to arrive at a single, representative value. To this end, Maturo

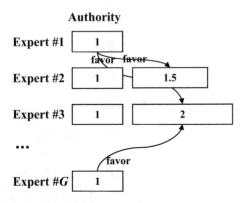

Fig. 2.7 Differentiating the authorities of experts

and Ventre [13] applied the weighted average method. In the view of Kuncheva and Krishnapuram [15], a consensus aggregator has to meet three requirements: symmetry, selective monotonicity, and unanimity. Three aggregation rules were proposed in their study as the minimum aggregation rule (e.g., FI), the maximum aggregation rule, and the average aggregation rule, for which the degree of consensus was measured in terms of the highest discrepancy [16].

2.6.1 Fuzzy Intersection

Fuzzy intersection or the minimum T-norm has been widely adopted to measure the consensus among experts and to aggregate fuzzy forecasts [5, 9]. Fuzzy intersection aggregates fuzzy forecasts as

$$\mu_{\tilde{I}(\{\tilde{y}_i(g)|g=1\sim G\})}(x) = \min\left(\left\{\mu_{\tilde{y}_i(g)}(x)\big|g=1\sim G\right\}\right) \qquad (2.50)$$

The aggregation result $\tilde{I}$ has a narrower range than the original fuzzy forecasts, while still containing the actual value. If all fuzzy yield forecasts are approximated with TFNs, then $\tilde{I}$ is a polygon-shaped fuzzy number, as illustrated in Fig. 2.8.

2.6.2 Partial-Consensus Fuzzy Intersection (PCFI)

When there is no overall consensus among all experts, the partial consensus among them, i.e., the consensus among most experts, can be sought instead.

Definition 1 (PCFI) [17] The g/G PCFI of the fuzzy forecasts by G experts, i.e., $\tilde{y}_i(1) \sim \tilde{y}_i(G)$, is indicated with $\widetilde{I}^{g/G}(\tilde{y}_i(1), \ldots, \tilde{y}_i(G))$ such that

Fig. 2.8 Fuzzy intersection of TFNs

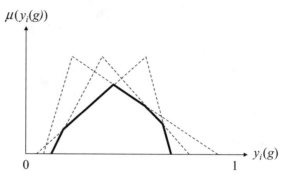

Fig. 2.9 PCFI result

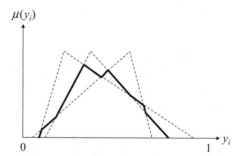

$$\mu_{\tilde{I}^{g/G}(\tilde{y}_i(1),\dots,\tilde{y}_i(G))}(x) = \max_{all\ h}\left(\min(\mu_{\tilde{y}_i(h(1))}(x),\dots,\mu_{\tilde{y}_i(h(g))}(x))\right) \quad (2.51)$$

where $h() \in \mathbf{Z}^+$; $1 \le h() \le G$; $h(p) \cap h(q) = \emptyset \; \forall \; p \ne q$; $g \ge 2$.

For example, the 2/3 PCFI of $\tilde{y}_i(1) \sim \tilde{y}_i(3)$ can be obtained as

$$\mu_{\tilde{I}^{g/G}(\tilde{y}_i(1),\dots,\tilde{y}_i(G))}(x) = \max_{all\ h}(\min(\mu_{\tilde{y}_i(1)}(x), \mu_{\tilde{y}_i(2)}(x)),$$

$$\min(\mu_{\tilde{y}_i(1)}(x), \mu_{\tilde{y}_i(3)}(x)), \min(\mu_{\tilde{y}_i(2)}(x), \mu_{\tilde{y}_i(3)}(x))) \quad (2.52)$$

which is illustrated in Fig. 2.9.

2.7 Defuzzification Mechanism

In practical applications, a crisp forecast is usually required. Therefore, a crisp forecast has to be generated from the polygon-shaped fuzzy forecast. For this purpose, many defuzzification methods (e.g., the center of gravity method, the mean-of-maxima method, the first-of-maxima method, the last-of-maxima method, the mean-of-support method, the first-of-support method, the last-of-support method, the height method, the mean-of-alpha-cut method, the basic defuzzification distribu-

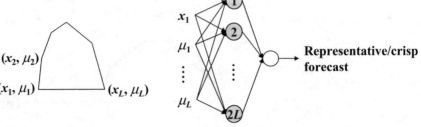

Fig. 2.10 Configuration of the BPN defuzzifier

tions method, etc.) are applicable [18]. The results of the mean-of-maxima method, the first-of-maxima method, and the last-of-maxima method are the same if the target is a TFN. After obtaining the defuzzified value, it is compared with the actual value to evaluate accuracy. However, among the existing defuzzification methods, no one method can surpass all the other methods in every case. Besides, the most suitable defuzzification method for a fuzzy variable is often chosen from the existing methods, which cannot guarantee the optimality of the chosen method. Further, the shape of the polygon-shaped fuzzy forecast is special. These phenomena provide a motive to propose a tailored defuzzification method. In Chen and Lin [9], a back-propagation network (BPN) was constructed for this purpose because theoretically a well-trained BPN (without being stuck to local minima) with a good selected topology could successfully map any complex distribution.

The configuration of the BPN defuzzifier is established as follows:

- Inputs: $2L$ parameters corresponding to the L corners of the polygon-shaped fuzzy forecast and the membership function values of these corners (see Fig. 2.10). Such a treatment is reasonable because the polygon-shaped fuzzy number obtained with the fuzzy intersection operator is a convex set, and therefore, every point in the polygon-shaped fuzzy number can be expressed with the combination of its corners. Besides, the results of many defuzzification methods can be represented with the corners of the fuzzy number (especially a TFN).
- Single hidden layer: Generally one or two hidden layers are more beneficial for the convergence property of the BPN.
- Number of neurons in the hidden layer, which is chosen from $1 \sim 2L$ according to a preliminary analysis, considering both effectiveness (the forecasting accuracy) and efficiency (the execution time).
- Output (o_i): the crisp forecast.
- Training algorithms: There are many advanced algorithms for training a BPN, e.g., the Fletcher–Reeves algorithm, the Broyden–Fletcher–Goldfarb–Shanno algorithm, the Levenberg–Marquardt algorithm, and the Bayesian regularization method [19].

The procedure for determining the parameter values is now described. Every polygon-shaped fuzzy forecast fed into the BPN is called an example. The number of examples is n. A portion of the examples is used as "training examples" into the BPN to determine the parameter values. Two phases are involved at the training stage. At first, in the forward phase, inputs are multiplied with weights, summated, and transferred to the hidden layer. Then, activated signals are outputted from the hidden layer and also transferred to the output layer with the same procedure. Finally, the output of the BPN is generated. After comparing the output with the actual value, the accuracy of the BPN, represented with root-mean-squared error (RMSE), can be evaluated. Subsequently in the backward phase, the Levenberg–Marquardt algorithm is recommended because it is much faster than the gradient descent algorithm for training the BPN [20]. The Levenberg–Marquardt algorithm was designed for training with second-order speed without having to compute the Hessian matrix. When training a BPN, the Hessian matrix can be approximated as

$$H = J^{\mathrm{T}} J \tag{2.53}$$

and the gradient can be computed as

$$g = J^{\mathrm{T}} e \tag{2.54}$$

where J is the Jacobian matrix containing the first derivatives of the network errors with respect to the weights and biases; e is the vector of the network errors. The Levenberg–Marquardt algorithm uses this approximation and updates the network parameters in a Newton-like way:

$$x_{t+1} = x_t - [J^{\mathrm{T}} J + \mu I]^{-1} J^{\mathrm{T}} e \tag{2.55}$$

where t is the epoch number; μ is Levenberg's damping factor. $x_{t+1} - x_t$ is the weight update vector that tells us by how much we should change our network weights to achieve a (possibly) better solution. When the scalar μ is zero, it is just Newton's method, using the approximated Hessian matrix. When μ is large, it is equivalent to the gradient descent algorithm with a small step size. Newton's method is faster and more accurate near an error minimum, so Levenberg–Marquardt algorithm's purpose is to move as quickly as possible to Newton's method. Thus, μ decreases after each successful step and increases only when a tentative step will increase the performance function. Consequently, the performance function is always reduced after each epoch.

In addition, if experts are of different authorities, then the corners formed by intersecting fuzzy forecasts with higher authorities will be learned more times in training the BPN.

2.8   Performance Evaluation in Fuzzy Collaborative Forecasting

Some performance measures for evaluating the precision and accuracy of a fuzzy forecasting method are defined as follows [21]:

- (Precision) The average range (AR):

$$AR = \frac{\sum_{i=1}^{n} (y_{i3} - y_{i1})}{n} \tag{2.56}$$

- (Precision) The hit rate (HR):

$$HR = \frac{\sum_{i=1}^{n} \xi_i}{n} \tag{2.57}$$

where

$$\xi_i = \begin{cases} 1 \text{ if } y_{i1} \leq y_i \leq y_{i3} \\ 0 \text{ otherwise} \end{cases} \tag{2.58}$$

- (Accuracy) Mean absolute error (MAE):

$$MAE = \frac{\sum_{i=1}^{n} |D(\tilde{y}_i) - y_i|}{n} \tag{2.59}$$

where $D()$ is the defuzzification function.
- (Accuracy) Mean absolute percentage error (MAPE):

$$MAPE = \frac{1}{n} \sum_{i=1}^{n} \frac{|D(\tilde{y}_i) - y_i|}{y_i} \cdot 100\% \tag{2.60}$$

- (Accuracy) Root-mean-squared error (RMSE):

$$RMSE = \sqrt{\frac{\sum_{i=1}^{n} (D(\tilde{y}_i) - y_i)^2}{n}} \tag{2.61}$$

All of these performance indicators, except HR, are as small as possible. Based on these definitions, the precision and accuracy of a fuzzy collaborative forecasting method can be assessed as follows:

$$AR = \frac{\sum_{i=1}^{n} (\max \tilde{I}_i - \min \tilde{I}_i)}{n} \tag{2.62}$$

$$HR = \frac{\sum_{i=1}^{n} \xi_i}{n} \tag{2.63}$$

where

$$\xi_i = \begin{cases} 1 \text{ if } \min \tilde{I}_i \leq y_i \leq \max \tilde{I}_i \\ 0 \text{ otherwise} \end{cases} \tag{2.64}$$

$$\mathrm{MAE} = \frac{\sum_{i=1}^{n} |a_i - y_i|}{n} \tag{2.65}$$

$$\mathrm{MAPE} = \frac{1}{n} \sum_{i=1}^{n} \frac{|a_i - y_i|}{y_i} \cdot 100\% \tag{2.66}$$

$$\mathrm{RMSE} = \sqrt{\frac{\sum_{i=1}^{n} (a_i - y_i)^2}{n}} \tag{2.67}$$

2.9 Quality of Collaboration Evaluation

Some ways to evaluate the quality of collaboration in fuzzy collaborative forecasting are defined as follows [21]:

- Maximum percentage improvement (MPI):

$$\mathrm{MPI}_{\mathrm{prec}} = \max_g \left(\frac{\mathrm{Prec}(\{\tilde{y}_i(g)\}) - \mathrm{Prec}(\mathrm{FCF})}{\mathrm{Prec}(\tilde{y}_i(g))} \right) \cdot 100\% \tag{2.68}$$

where $\mathrm{Prec}(\{\tilde{y}_i(g)\})$ is the forecasting precision achieved by expert g, while $\mathrm{Prec}(FCF)$ is the precision achieved using the fuzzy collaborative forecasting approach. Therefore, $\mathrm{Prec}() = \mathrm{AR}()$ or $\mathrm{HR}()$.

$$\mathrm{MPI}_{\mathrm{accur}} = \max_g \left(\frac{\mathrm{Accu}(\{\tilde{y}_i(g)\}) - \mathrm{Accu}(\mathrm{FCF})}{\mathrm{Accu}(\tilde{y}_i(g))} \right) \cdot 100\% \tag{2.69}$$

where $\mathrm{Accu}(\{\tilde{y}_i(g)\})$ is the forecasting accuracy achieved by expert g, while $\mathrm{Accu}(\mathrm{FCF})$ is the accuracy achieved using the fuzzy collaborative forecasting approach. Therefore, $\mathrm{Accu}() = \mathrm{MAE}()$, $\mathrm{MAPE}()$, or $\mathrm{RMSE}()$.

- Average percentage improvement (API):

$$\mathrm{API}_{\mathrm{prec}} = \frac{1}{G} \sum_{g=1}^{G} \left(\frac{\mathrm{Prec}(\{\tilde{y}_i(g)\}) - \mathrm{Prec}(\mathrm{FCF})}{\mathrm{Prec}(\tilde{y}_i(g))} \right) \cdot 100\% \tag{2.70}$$

$$\mathrm{API}_{\mathrm{accu}} = \frac{1}{G} \sum_{g=1}^{G} \left(\frac{\mathrm{Accu}(\{\tilde{y}_i(g)\}) - \mathrm{Accu}(\mathrm{FCF})}{\mathrm{Accu}(\tilde{y}_i(g))} \right) \cdot 100\% \tag{2.71}$$

References

1. K.B. Song, Y.S. Baek, D.H. Hong, G. Jang, Short-term load forecasting for the holidays using fuzzy linear regression method. IEEE Trans. Power Syst. **20**(1), 96–101 (2005)
2. J. Watada, H. Tanaka, T. Shimomura, Identification of learning curve based on possibilistic concepts. Adv. Human Factors/Ergon. **6**, 191–208 (1986)
3. T. Chen, M.J. Wang, A fuzzy set approach for yield learning modeling in wafer manufacturing. IEEE Trans. Semicond. Manuf. **12**(2), 252–258 (1999)
4. F.M. Tseng, G.H. Tzeng, H.C. Yu, B.J. Yuan, Fuzzy ARIMA model for forecasting the foreign exchange market. Fuzzy Sets Syst. **118**(1), 9–19 (2001)
5. T. Chen, Y.C. Wang, An agent-based fuzzy collaborative intelligence approach for precise and accurate semiconductor yield forecasting. IEEE Trans. Fuzzy Syst. **22**(1), 201–211 (2014)
6. H. Tanaka, J. Watada, Possibilistic linear systems and their application to the linear regression model. Fuzzy Sets Syst. **27**(3), 275–289 (1988)
7. G. Peters, Fuzzy linear regression with fuzzy intervals. Fuzzy Sets Syst. **63**(1), 45–55 (1994)
8. S. Donoso, N. Marin, M.A. Vila, Quadratic programming models for fuzzy regression, in *Proceedings of International Conference on Mathematical and Statistical Modeling in Honor of Enrique Castillo* (2006)
9. T. Chen, Y.C. Lin, A fuzzy-neural system incorporating unequally important expert opinions for semiconductor yield forecasting. Int. J. Uncertainty Fuzziness Knowledge-Based Syst. **16**(01), 35–58 (2008)
10. T. Chen, An innovative fuzzy and artificial neural network approach for forecasting yield under an uncertain learning environment. J. Ambient Intell. Humanized Comput. (2018)
11. J. Nocedal, S. Wright, *Numerical Optimization* (Springer Science & Business Media, New York, 2006)
12. I.S. Cheng, Y. Tsujimura, M. Gen, T. Tozawa, An efficient approach for large scale project planning based on fuzzy Delphi method. Fuzzy Sets Syst. **76**, 277–288 (1995)
13. A. Maturo, A.G.S. Ventre, Models for consensus in multiperson decision making, in *2008 Annual Meeting of the North American Fuzzy Information Processing Society* (2008), pp. 1–4
14. T. Chen, An online collaborative semiconductor yield forecasting system. Expert Syst. Appl. **36**(3), 5830–5843 (2009)
15. L.I. Kuncheva, R. Krishnapuram, A fuzzy consensus aggregation operator. Fuzzy Sets Syst. **79**, 347–356 (1996)
16. Y.C. Wang, T. Chen, A partial-consensus posterior-aggregation FAHP method—supplier selection problem as an example. Mathematics **7**(2), 179 (2019)
17. T. Chen, A collaborative fuzzy-neural system for global CO_2 concentration forecasting. Int. J. Innov. Comput. Inf. Control **8**(11), 7679–7696 (2012)
18. X. Liu, Parameterized defuzzification with maximum entropy weighting function—another view of the weighting function expectation method. Math. Comput. Model. **45**, 177–188 (2007)
19. E. Eraslan, The estimation of product standard time by artificial neural networks in the molding industry. Math. Probl. Eng. article ID 527452 (2009)
20. A. Ranganathan, The Levenberg-Marquardt Algorithm (2004). Available: http://www.scribd.com/doc/10093320/Levenberg-Marquardt-Algorithm
21. T. Chen, Forecasting the unit cost of a product with some linear fuzzy collaborative forecasting models. Algorithms **5**(4), 449–468 (2012)

Chapter 3
Nonlinear Fuzzy Collaborative Forecasting Methods

3.1 Nonlinear Fuzzy Forecasting Methods

Various nonlinear fuzzy methods have been applied to forecasting. For example, fuzzy inference systems (FISs), such as Mamdani FISs, Sugeno [or Takagi-Sugeno-Kang (TSK)] FISs and Tsukamoto FISs, are actually nonlinear fuzzy methods that have been extensively applied to short-term load (i.e., electricity demand) [1], real-time flood (in terms of river flows) [2], stock price [3], daily water consumption [4], and short-term power price forecasting [5], etc. The implication, aggregation, and defuzzification operators used in the three types of FISs are different, as compared in Table 3.1. These FISs are easy to understand and communicate and implement using existing software packages. However, a fuzzy forecast generated by the FISs may not include the actual value, even for the training data.

The incorporation of FIS and artificial neural network (ANN) results in an adaptive neuro-fuzzy inference system (ANFIS) that has been applied to predict reservoir water level [6], monthly stock market return [7], rock elastic constant [8], building energy needs [9], etc. An ANFIS enumerates and aggregates possible combinations of fuzzy inference rules to optimize the forecasting performance. The five layers of an ANFIS correspond to the five steps of a FIS. Fuzzy neural networks (FNN) constructed by fuzzifying ANNs have also been proposed for forecasting purposes.

This chapter is composed of two parts. In the first part, some nonlinear fuzzy forecasting methods for generating a fuzzy forecast that contains the actual value are reviewed. A modified back-propagation network (BPN) approach is introduced in Sect. 3.2. In Sect. 3.3, a fuzzy back-propagation network (FBPN) approach is introduced. In Sect. 3.4, a simplified calculation technique is applied to simplify the required calculation in the FBPN approach.

T.-C. T. Chen and K. Honda, *Fuzzy Collaborative Forecasting and Clustering*, SpringerBriefs in Applied Sciences and Technology, https://doi.org/10.1007/978-3-030-22574-2_3

Table 3.1 Implication, aggregation, and defuzzification operators used in the three types of FISs

FIS	Implication operator	Aggregation operator	Defuzzification operator
Mamdani	Min	Max	Center of gravity
TSK	Product	Weighted average	–
Tsukamoto	Min	Max	Height defuzzification

In the second part, some nonlinear fuzzy collaborative forecasting methods based on such fuzzy forecasts are introduced, including a fuzzy collaborative forecasting method based on FBPNs, and a collaborative fuzzy analytic hierarchy process approach. Without the loss of generality, all parameters in the methods are given in or approximated with triangular fuzzy numbers (TFNs).

3.2 A Modified Back-Propagation Network (BPN) Approach for Generating Fuzzy Forecasts

Chen and Wang [10] introduced a modified BPN approach for generating a fuzzy forecast that includes the actual value (for the training data). Although there have been some more advanced artificial neural networks, such as a compositional pattern-producing network, cascading neural network, or dynamic neural network, a well-trained BPN with an optimized structure can still produce very good results [11]. The configuration of the BPN is as follows:

- Inputs: m inputs for example i, indicated with $x_{ij}, j = 1 \sim m$, that are used to forecast the target y_i. To facilitate the search for solutions, it is strongly recommended to normalize the inputs to a range narrower than [0 1] [12]:

$$N(x_{ij}) = N_L + \frac{x_{ij} - \min_p x_{pj}}{\max_p x_{pj} - \min_p x_{pj}} (N_U - N_L) \qquad (3.1)$$

where $N(x_{ij})$ is the normalized value of x_{ij}; N_L and N_U indicate the lower and upper bounds of the range of the normalized value, respectively. $\min_p x_{pj}$ and $\max_p x_{pj}$ are the minimum and maximum of x_{pj}, respectively. The formula can be written as

$$x_{ij} = \frac{N(x_{ij}) - N_L}{N_U - N_L} (\max_p x_{pj} - \min_p x_{pj}) + \min_p x_{pj} \qquad (3.2)$$

if the un-normalized value is to be obtained instead.

- The BPN has only one hidden layer. Two or more hidden layers slow down the convergence speed and may not lead to any better solution. The number of nodes in the hidden layer is chosen from 1 to 2 m after trying each of them.
- The output from the BPN is the normalized value of the core of the fuzzy forecast, i.e., $o_i = N(y_{i2})$, that can be compared with the normalized value of the actual value, i.e., $N(y_i)$.
- The activation function used for the hidden layer is the hyperbolic tangent sigmoid function, and for the others is the linear activation function.
- T epochs will be run each time. The start conditions will be randomized to reduce the possibility of being stuck on local optima.

The network parameters are defined as follows:

- w_{jk}^h: the weight of the connection between input node j and hidden-layer node k.
- w_l^o: the weight of the connection between hidden-layer node l and the output node.
- θ_k^h: the threshold for screening out weak signals by hidden-layer node k.
- θ^o: the threshold for screening out weak signals by the output node.

The procedure for determining the parameter values is now described. A portion of examples is fed as "training examples" into the BPN to determine the parameter values. Two phases are involved at the training stage. At first, in the forward phase, inputs are multiplied with weights, summated, and transferred to the hidden layer. Then activated signals are outputted from the hidden layer as:

$$h_{ik} = \frac{1}{1 + e^{-n_{ik}^h}} \tag{3.3}$$

where

$$n_{ik}^h = I_{ik}^h - \theta_k^h \tag{3.4}$$

$$I_{ik}^h = \sum_j (w_{jk}^h \cdot x_{ij}) \tag{3.5}$$

h_{ik}s are also transferred to the output layer with the same procedure. Finally, the output of the BPN is generated as:

$$o_i = \frac{1}{1 + e^{-n_i^o}} \tag{3.6}$$

where

$$n_i^o = I_i^o - \theta^o \tag{3.7}$$

$$I_i^o = \sum_j (w_k^o \cdot h_{ik}) \tag{3.8}$$

Some algorithms are applicable for training a BPN, such as the gradient descent (GD) algorithm, the conjugate gradient algorithms, the Levenberg-Marquardt (LM) algorithm, and others [13]. After training, the optimized network parameters are indicated with w_{jk}^{h*}, w_l^{o*}, θ_k^{h*}, and θ^{o*}.

Subsequently, the parameter values in the BPN are adjusted to determine the normalized value of the upper bound of the fuzzy forecast, i.e., $o_i' = N(y_{i3})$. Assume the adjustments made to the parameter values are indicated with Δw_{jk}^h, Δw_l^o, $\Delta \theta_k^h$, and $\Delta \theta^o$, respectively. After adjustment, it is expected that

$$o_i' \geq N(y_i) \tag{3.9}$$

where

$$o_i' = \frac{1}{1 + e^{-n_i^{o'}}} \tag{3.10}$$

$$n_i^{o'} = I_i^{o'} - \theta^{o'} = I_i^{o'} - (\theta^{o*} + \Delta \theta^o) \tag{3.11}$$

$$I_i^{o'} = \sum_j (w_k^{o'} \cdot h_{ik}') = \sum_j ((w_k^{o*} + \Delta w_k^o) \cdot h_{ik}') \tag{3.12}$$

in which the outputs from the hidden-layer nodes are equal to

$$h_{ik}' = \frac{1}{1 + e^{-n_{ik}^{h'}}} \tag{3.13}$$

where

$$n_{ik}^{h'} = I_{ik}^{h'} - \theta_k^{h'} = I_{ik}^{h'} - \left(\theta_k^{h*} + \Delta \theta_k^h\right) \tag{3.14}$$

$$I_{ik}^{h'} = \sum_j \left(w_{jk}^{h'} \cdot x_{ij}\right) = \sum_j \left((w_{jk}^{h*} + \Delta w_{jk}^h) \cdot x_{ij}\right) \tag{3.15}$$

Substituting these equations into the inequality and the following result is generated:

$$-\sum_j \left((w_k^{o*} + \Delta w_k^o) \cdot h_{ik}'\right) \leq \ln\left(\frac{1}{N(y_i)} - 1\right) - \left(\theta^{o*} + \Delta \theta^o\right) \tag{3.16}$$

which is a nonlinear constraint. On the other hand, the objective function is to minimize the sum of the mid-ranges on the right-hand sides of the fuzzy forecasts:

$$\text{Min } Z_1 = \sum_i \left(o_i' - o_i^*\right) \tag{3.17}$$

where o_i^* is a constant, and therefore the objective function becomes

$$\text{Min } Z_1 = \sum_i o_i' \tag{3.18}$$

which after expansion becomes

$$\text{Min } Z_1 = \sum_i \frac{1}{1 + e^{-\left(\sum_j \left((w_k^{o*} + \Delta w_k^o) \cdot h_{ik}'\right) - (\theta^{o*} + \Delta\theta^o)\right)}} \tag{3.19}$$

Finally, the following nonlinear programing (NLP) model is constructed:

(Model NLP I)

$$\text{Min } Z_1 = \sum_i \frac{1}{1 + e^{-(\sum_j ((w_k^{o*} + \Delta w_k^o) \cdot h_{ik}') - (\theta^{o*} + \Delta\theta^o))}} \tag{3.20}$$

subject to

$$-\sum_j \left((w_k^{o*} + \Delta w_k^o) \cdot h_{ik}'\right) \leq \ln\left(\frac{1}{N(y_i)} - 1\right) - (\theta^{o*} + \Delta\theta^o); \quad i = 1 \sim n \tag{3.21}$$

$$h_{ik}' = \frac{1}{1 + e^{-\left(\sum_j \left((w_{jk}^{h*} + \Delta w_{jk}^h) \cdot x_{ij}\right) - (\theta_k^{h*} + \Delta\theta_k^h)\right)}}; \quad i = 1 \sim n; \ k = 1 \sim K \tag{3.22}$$

$$\Delta w_{jk}^h, \Delta w_l^o, \Delta\theta_k^h, \Delta\theta^o \in \mathbf{R}; \quad i = 1 \sim n; \ j = 1 \sim m; \ k = 1 \sim K \tag{3.23}$$

In a similar way, the parameter values in the BPN can be adjusted to determine the normalized value of the lower bound of the fuzzy forecast, i.e., $o_i' = N(y_{i1})$, which relies on the following NLP model:

(Model NLP II)

$$\text{Max } Z_2 = \sum_i \frac{1}{1 + e^{-\left(\sum_j \left((w_k^{o*} + \Delta w_k^o) \cdot h_{ik}'\right) - (\theta^{o*} + \Delta\theta^o)\right)}} \tag{3.24}$$

subject to

$$-\sum_j \left((w_k^{o*} + \Delta w_k^o) \cdot h_{ik}'\right) \geq \ln\left(\frac{1}{N(y_i)} - 1\right) - (\theta^{o*} + \Delta\theta^o); \quad i = 1 \sim n \tag{3.25}$$

$$h_{ik}' = \frac{1}{1 + e^{-\left(\sum_j \left((w_{jk}^{h*} + \Delta w_{jk}^h) \cdot x_{ij}\right) - (\theta_k^{h*} + \Delta\theta_k^h)\right)}}; \quad i = 1 \sim n; \ k = 1 \sim K \tag{3.26}$$

$$\Delta w^h_{jk}, \Delta w^o_l, \Delta \theta^h_k, \Delta \theta^o \in \mathbf{R}; \quad i = 1 \sim n; \quad j = 1 \sim m; \quad k = 1 \sim K \qquad (3.27)$$

The two NLP problems are not easy to solve. To address this issue, Chen and Wang [10] applied the concept of goal programing (GP) to help solve the NLP problems. For example, in model NLP I, at first, after expansion h'_{ik} becomes

$$-\left(\sum_j \left(\left(w^{h*}_{jk} + \Delta w^h_{jk} \right) \cdot x_{ij} \right) - \left(\theta^{h*}_k + \Delta \theta^h_k \right) \right) = \ln \left(\frac{1}{h'_{ik}} - 1 \right) \qquad (3.28)$$

An upper bound ξ_{ik} can be established for h'_{ik}:

$$h'_{ik} \le \xi_{ik} \qquad (3.29)$$

Therefore,

$$\ln \left(\frac{1}{h'_{ik}} - 1 \right) \ge \ln \left(\frac{1}{\xi_{ik}} - 1 \right) \qquad (3.30)$$

Substituting (3.30) into (3.28) gives

$$-\left(\sum_j \left(\left(w^{h*}_{jk} + \Delta w^h_{jk} \right) \cdot x_{ij} \right) - \left(\theta^{h*}_k + \Delta \theta^h_k \right) \right) \ge \ln \left(\frac{1}{\xi_{ik}} - 1 \right) \qquad (3.31)$$

Subsequently, an upper bound ψ_i can be established for o'_i such that $o'_i \le \psi_i$:

$$\frac{1}{1 + e^{-\left(\sum_j \left(\left(w^{o*}_k + \Delta w^o_k \right) \cdot h'_{ik} \right) - \left(\theta^{o*} + \Delta \theta^o \right) \right)}} \le \psi_i \qquad (3.32)$$

which after expansion becomes

$$-\sum_j \left(\left(w^{o*}_k + \Delta w^o_k \right) \cdot h'_{ik} \right) \le \ln \left(\frac{1}{\psi_i} - 1 \right) - \left(\theta^{o*} + \Delta \theta^o \right) \qquad (3.33)$$

which is a new linear constraint. The new objective function is

$$\text{Min } Z_3 = \sum_i \psi_i \qquad (3.34)$$

which is linear. ψ_i is an established positive goal. Various values of ψ_i will be fed into the problem, and therefore the problem is solved many times. From these optimization results, the best one is adopted.

3.3 A Fuzzy Back-Propagation Network (FBPN) Approach

A FBPN can be constructed to forecast the target to insure a 100% inclusion level. A FBPN is a fuzzified BPN. The configuration of the FBPN is the same as that of the BPN. The procedure for determining the parameter values is now described. A portion of examples is fed as "training examples" into the FBPN to determine the parameter values. Two phases are involved at the training stage. At first, in the forward phase, inputs are multiplied with weights, summated, and transferred to the hidden layer. Then activated signals are outputted from the hidden layer as:

$$\tilde{h}_{ik} = (h_{ik1}, h_{ik2}, h_{ik3}) = \frac{1}{1 + e^{-\tilde{n}_{ik}^h}} = \left(\frac{1}{1 + e^{-n_{ik1}^h}}, \frac{1}{1 + e^{-n_{ik2}^h}}, \frac{1}{1 + e^{-n_{ik3}^h}} \right)$$

$$(3.35)$$

where

$$\tilde{n}_{ik}^h = \left(n_{ik1}^h, n_{ik2}^h, n_{ik3}^h \right) = \tilde{I}_{ik}^h (-) \tilde{\theta}_k^h = \left(I_{ik1}^h - \theta_{k3}^h, I_{ik2}^h - \theta_{k2}^h, I_{ik3}^h - \theta_{k1}^h \right) \quad (3.36)$$

$$\tilde{I}_{ik}^h = \left(I_{ik1}^h, I_{ik2}^h, I_{ik3}^h \right) = \sum_j \tilde{w}_{jk}^h \cdot x_{ij}$$

$$= \left(\sum_j \min\left(w_{jk1}^h x_{ij}, w_{jk3}^h x_{ij} \right), w_{jk2}^h x_{ij}, \sum_j \max\left(w_{jk1}^h x_{ij}, w_{jk3}^h x_{ij} \right) \right) \quad (3.37)$$

$(-)$ denotes fuzzy subtraction; $\tilde{h}_{ik}$s are also transferred to the output layer with the same procedure. Finally, the output of the FBPN is generated as:

$$\tilde{o}_i = (o_{i1}, o_{i2}, o_{i3}) = \frac{1}{1 + e^{-\tilde{n}_i^o}} = \left(\frac{1}{1 + e^{-n_{i1}^o}}, \frac{1}{1 + e^{-n_{i2}^o}}, \frac{1}{1 + e^{-n_{i3}^o}} \right) \quad (3.38)$$

where

$$\tilde{n}_i^o = \left(n_{i1}^o, n_{i2}^o, n_{i3}^o \right) = \tilde{I}_i^o (-) \tilde{\theta}^o = \left(I_{i1}^o - \theta_3^o, I_{i2}^o - \theta_2^o, I_{i3}^o - \theta_1^o \right) \quad (3.39)$$

$$\tilde{I}_i^o = \left(I_{i1}^o, I_{i2}^o, I_{i3}^o \right) = \sum_j \tilde{w}_k^o (\times) \tilde{h}_{ik}$$

$$= \left(\sum_j \min\left(w_{k1}^o h_{ik1}, w_{k3}^o h_{ik3} \right), w_{k2}^o h_{ik2}, \sum_j \max\left(w_{k1}^o h_{ik1}, w_{k3}^o h_{ik3} \right) \right) \quad (3.40)$$

Subsequently in the backward phase, the training of the FBPN is decomposed into three subtasks: determining the core value, upper, and lower bounds of the parameters.

First, to determine the core of each fuzzy parameter (such as w_{jk2}^h, θ_{k2}^h, w_{k2}^o, and θ_2^o), the FBPN is treated as a crisp one and trained using algorithms such as the GD algorithm, the LM algorithm, the conjugate gradient algorithms, etc.

Then, the following NLP problem is solved to determine the lower and upper bounds of each fuzzy parameter derive the values of network parameters:

(Model NLP III)

$$\text{Min } Z_5 = \sum_i (o_{i3} - o_{i1}) \tag{3.41}$$

subject to

$$o_{i3} \geq o_{i2}^* \tag{3.42}$$

$$o_{i3} \geq N(y_i) \tag{3.43}$$

$$o_{i1} \leq o_{i2}^* \tag{3.44}$$

$$o_{i1} \leq N(y_i) \tag{3.45}$$

$$-\left(\sum_j \min\left(w_{k1}^o h_{ik1}, w_{k3}^o h_{ik3}\right) - \theta_3^o\right) = \ln\left(\frac{1}{o_{i1}} - 1\right) \tag{3.46}$$

$$-\left(\sum_j \max\left(w_{k1}^o h_{ik1}, w_{k3}^o h_{ik3}\right) - \theta_1^o\right) = \ln\left(\frac{1}{o_{i3}} - 1\right) \tag{3.47}$$

$$-\left(\sum_j \min\left(w_{jk1}^h x_{ij}, w_{jk3}^h x_{ij}\right) - \theta_{k3}^h\right) = \ln\left(\frac{1}{h_{ik1}} - 1\right) \tag{3.48}$$

$$-\left(\sum_j \max\left(w_{jk1}^h x_{ij}, w_{jk3}^h x_{ij}\right) - \theta_{k1}^h\right) = \ln\left(\frac{1}{h_{ik3}} - 1\right) \tag{3.49}$$

$$w_{jk1}^h \leq w_{jk2}^{h*} \leq w_{jk3}^h; \quad j = 1 \sim M \tag{3.50}$$

$$w_{k1}^o \leq w_{k2}^{o*} \leq w_{k3}^o; \quad k = 1 \sim K \tag{3.51}$$

$$\theta_{k1}^h \leq \theta_{k2}^{h*} \leq \theta_{k3}^h; \quad k = 1 \sim K \tag{3.52}$$

$$\theta_1^o \leq \theta_2^{o*} \leq \theta_3^o \tag{3.53}$$

3.4 A Simplified Calculation Technique

To simplify the required calculation, Chen [14, 15] proposed a simplified calculation technique for establishing the upper and lower bounds of a fuzzy forecast. Among the BPN parameters established, only the threshold on the output node was modified, simplifying the procedure and enabling it to yield satisfactory results:

$$w^h_{jk1} = w^{h*}_{jk2} = w^h_{jk3}; \quad j = 1 \sim M \tag{3.54}$$

$$w^o_{k1} = w^{o*}_{k2} = w^o_{k3}; \quad k = 1 \sim K \tag{3.55}$$

$$\theta^h_{k1} = \theta^{h*}_{k2} = \theta^h_{k3}; \quad k = 1 \sim K \tag{3.56}$$

Chen also derived two equations that are independent of BPN parameters for the same purpose:

$$\theta^o_3 = \theta^{o*}_2 + \min_i \left(\ln\left(\frac{1}{N(y_i)} - 1\right) - \ln\left(\frac{1}{o^*_{i2}} - 1\right) \right) \tag{3.57}$$

$$\theta^o_1 = \theta^{o*}_2 - \max_i \left(\ln\left(\frac{1}{N(y_i)} - 1\right) - \ln\left(\frac{1}{o^*_{i2}} - 1\right) \right) \tag{3.58}$$

3.5 A Fuzzy Collaborative Forecasting Method Based on Fuzzy Back-propagation Networks (FBPNs)

In the FCF method based on FBPNs, a group composed of a number of experts (or software agents) is formed. These experts hold different points of view on the following aspects:

(1) The right mid-range of the fuzzy forecast ($\psi_i(g)$).
(2) The left mid-range of the fuzzy forecast ($\pi_i(g)$).
(3) The extent that a fuzzy forecast is satisfied with the actual value on the right-hand side ($s_U(g)$) (see Fig. 3.1a).
(4) The extent that a fuzzy forecast is satisfied with the actual value on the left-hand side ($s_L(g)$) (see Fig. 3.1b).

These views are incorporated into the FBPN approach to generate fuzzy-valued forecasts. First, the right mid-range of the fuzzy forecast is narrower than $\psi_i(g)$:

$$o_{i3}(g) - o^*_{i2}(g) \le \psi_i(g) \tag{3.59}$$

while the left mid-range of the fuzzy forecast has to be narrower than $\pi_i(g)$:

$$o^*_{i2}(g) - o_{i1}(g) \le \pi_i(g) \tag{3.60}$$

Fig. 3.1 Point of view of an expert

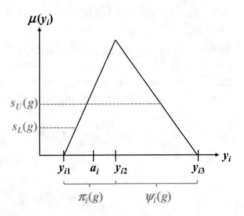

Subsequently, the extent that a fuzzy forecast is satisfied with the actual value on the right-hand side is higher than $s_U(g)$:

$$\frac{y - o_{i3}(g)}{o_{i2}^*(g) - o_{i3}(g)} \geq s_U(g) \tag{3.61}$$

or

$$y_i \leq s_U(g)o_{i2}^*(g) + (1 - s_U(g))o_{i3}(g) \tag{3.62}$$

while the extent that a fuzzy forecast is satisfied with the actual value on the left-hand side has to be higher than $s_L(g)$:

$$\frac{y - o_{i1}(g)}{o_{i2}^*(g) - o_{i1}(g)} \geq s_L(g) \tag{3.63}$$

or

$$y_i \geq s_L(g)o_{i2}^*(g) + (1 - s_L(g))o_{i1}(g) \tag{3.64}$$

These linear constraints are added to Model NLP III as additional constraints. Obviously, the values of these parameters set by different experts may not be equal, which results in different fuzzy forecasts that need to be aggregated. In addition, the forecasting results by all experts can be communicated to each other, so that they can modify their settings, and generate closer fuzzy forecasts.

3.6 A Collaborative Fuzzy Analytic Hierarchy Process (FAHP) Approach

Collaborative fuzzy analytic hierarchy process (FAHP) is a special application of nonlinear fuzzy collaborative forecasting, in which the priorities (or weights) of factors (attributes, or criteria) that are acceptable to all decision makers (DMs) are to be estimated. In other words, a multi-DM post-aggregation FAHP problem is analogous to an unsupervised fuzzy collaborative forecasting problem, since there are no actual values of the fuzzy priorities.

FAHP is a variant of the analytic hierarchy process (AHP) that has been widely applied to multi-criteria decision-making problems in various fields [16–18]. AHP is based on pair-wise comparison results that are subjective. To address this issue, AHP usually aggregates the pair-wise comparison results by multiple DMs [19–21]. However, a DM may be uncertain whether the pair-wise comparison results are reflective of his/her beliefs or not. To consider such uncertainty, pair-wise comparison results can be mapped to fuzzy values. The two treatments give rise to the prevalent multi-DM FAHP methods.

A collaborative FAHP approach is introduced in this section for comparing the relative priorities of several factors. The collaborative FAHP approach consists of the following steps [22]:

- Step 1. Each DM applies the fuzzy geometric mean (FGM) approach to estimate the fuzzy priorities.
- Step 2. Apply fuzzy intersection (FI) to aggregate the estimation results, so as to derive the narrowest range of each fuzzy priority.
- Step 3. If the overall consensus among the DMs does not exist, go to Step 4; otherwise, go to Step 5.
- Step 4. Apply partial-consensus fuzzy intersection (PCFI) to aggregate the estimation results, so as to derive the narrowest range of each fuzzy priority.
- Step 5. Apply COG to defuzzify the aggregation result, so as to generate a crisp/representative value.

The procedure of the collaborative FAHP approach is illustrated with a flowchart in Fig. 3.2.

3.6.1 Fuzzy Geometric Mean (FGM) for Estimating the Fuzzy Priorities

In the FGM approach, at first each DM expresses his/her opinion on the relative priority of a factor over that of another with linguistic terms such as "as equal as," "weakly more important than," "strongly more important than," "very strongly more important than," and "absolutely more important than." Without the loss of generality, these linguistic terms can be mapped to TFNs such as:

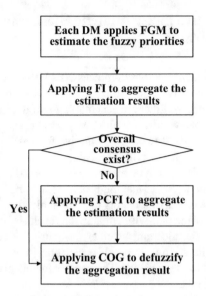

Fig. 3.2 Procedure of the collaborative FAHP approach

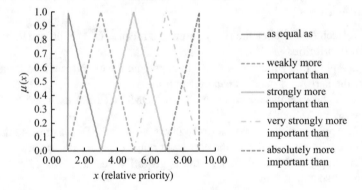

Fig. 3.3 TFNs for the linguistic terms

- L1: "As equal as" = (1, 1, 3);
- L2: "Weakly more important than" = (1, 3, 5);
- L3: "Strongly more important than" = (3, 5, 7);
- L4: "Very strongly more important than" = (5, 7, 9);
- L5: "Absolutely more important than" = (7, 9, 9);

which are illustrated in Fig. 3.3. It is a theoretically challenging task to find a set of TFNs for the linguistic terms to increase the possibility of reaching a consensus.

Based on the pair-wise comparison results by the g-th DM, a fuzzy pair-wise comparison matrix is constructed as:

$$\tilde{\mathbf{A}}_{n \times n}(g) = [\tilde{a}_{ij}(g)]; \quad i, j = 1 \sim n \tag{3.65}$$

where

$$\tilde{a}_{ij}(g) = \begin{cases} 1 & \text{if } i = j \\ \frac{1}{\tilde{a}_{ji}(g)} & \text{otherwise} \end{cases}; \quad i, j = 1 \sim n \tag{3.66}$$

$\tilde{a}_{ij}(g)$ is the fuzzy pair-wise comparison result embodying the relative priority of factor i over factor j. $\tilde{a}_{ij}(g)$ is chosen from the linguistic terms in Fig. 3.3. $\tilde{a}_{ij}(g)$ is a positive comparison if $\tilde{a}_{ij}(g) \geq 1$. The fuzzy eigenvalue and eigenvector of $\tilde{\mathbf{A}}(g)$, indicated, respectively, with $\tilde{\lambda}(g)$ and $\tilde{\mathbf{x}}(g)$, satisfy

$$\det(\tilde{\mathbf{A}}(g)(-)\tilde{\lambda}(g)\mathbf{I}) = 0 \tag{3.67}$$

and

$$(\tilde{\mathbf{A}}(g)(-)\tilde{\lambda}(g)\mathbf{I})(\times)\tilde{\mathbf{x}}(g) = 0 \tag{3.68}$$

where $(-)$ and $(\times)$ denote fuzzy subtraction and multiplication, respectively. The fuzzy maximal eigenvalue and the fuzzy priority of each criterion are derived, respectively, as

$$\tilde{\lambda}_{\max}(g) = \max\tilde{\lambda}(g) \tag{3.69}$$

$$\tilde{w}_i(g) = \frac{\tilde{x}_i(g)}{\sum_{j=1}^{n} \tilde{x}_j(g)} \tag{3.70}$$

The FGM method estimates the values of fuzzy priorities as

$$\tilde{\mathbf{w}}(g) = [\tilde{w}_i(g)]^{\mathrm{T}} = \left[\frac{\sqrt[n]{\prod_{j=1}^{n} \tilde{a}_{ij}(g)}}{\sum_{i=1}^{n} \sqrt[n]{\prod_{j=1}^{n} \tilde{a}_{ij}(g)}} \right]^{\mathrm{T}} \tag{3.71}$$

Based on (3.71), the fuzzy maximal eigenvalue can be estimated as

$$\tilde{\lambda}_{\max}(g) = \frac{\tilde{\mathbf{A}}(g)(\times)\tilde{\mathbf{w}}(g)}{\tilde{\mathbf{w}}(g)}$$
$$= \frac{1}{n} \sum_{i=1}^{n} \left(\frac{\sum_{j=1}^{n} (\tilde{a}_{ij}(g)(\times)\tilde{w}_i(g))}{\tilde{w}_i(g)} \right) \tag{3.72}$$

Obviously, Eqs. (3.71) and (3.72) are fuzzy weighted average (FWA) problems. A variant of FGM (a simplified yet more prevalent version), indicated with FGMi, is to ignore the dependency between the dividend and divisor of either equation. Based on

$\tilde{\lambda}_{max}(g)$, the consistency among the fuzzy pair-wise comparison results is evaluated
as

$$\text{Consistency index}: \widetilde{\text{C.I.}}(g) = \frac{\tilde{\lambda}_{max}(g) - n}{n - 1} \qquad (3.73)$$

$$\text{Consistency ratio}: \widetilde{\text{C.R.}}(g) = \frac{\widetilde{\text{C.I.}}(g)}{\text{R.I.}} \qquad (3.74)$$

where *RI* is the random index [23]. The fuzzy pair-wise comparison results are incon-
sistent if $\widetilde{\text{C.I.}}(g) < 0.1$ or $\widetilde{\text{C.R.}}(g) < 0.1$ [23], which can be relaxed to $\widetilde{\text{C.I.}}(g) < 0.3$
or $\widetilde{\text{C.R.}}(g) < 0.3$ if the matrix size is large [24].

3.6.2   Finding Out the Overall Consensus Using Fuzzy Intersection (FI)

As mentioned previously, a multi-DM post-aggregation FAHP problem is analogous to an unsupervised FCF problem to which a suitable consensus aggregator is
critical. There are many possible mechanisms applicable for this purpose, e.g., the
fuzzy weighted average [25, 26], a fuzzy back-propagation network [27], the fuzzy
intersection/AND operator [28], the fuzzy union/OR operator [28], and others [29,
30]. The advantages and/or disadvantages of these mechanisms are compared in
Table 3.2.

Remark If the fuzzy forecasts made by all experts contain the actual value, then the
aggregation result using each of the mechanisms also contains the actual value.

Table 3.2 Advantages and/or disadvantages of various collaboration and aggregation mechanisms

Aggregation mechanism	Easy to communicate	Easy to implement	Execution time	Effect on improving the precision	Effect on improving accuracy
Fuzzy weighted average	Very easy	Very easy	Very short	Very insignificant	Case by case
Fuzzy back-propagation network	Very difficult	Very difficult	Very lengthy	Insignificant	Very significant
Fuzzy union	Easy	Easy	Short	Very insignificant	Case by case
Fuzzy intersection	Easy	Easy	Short	Very significant	Case by case

Among them, FI is the most prevalent consensus aggregator in fuzzy collaborative forecasting studies [31]. FI finds out the values common to those estimated by all DMs. Therefore, it can be used to find out the overall consensus among the DMs. When the fuzzy priority estimated by each DM is approximated with a TFN, the FI result will be a polygon-shaped fuzzy number (refer to Fig. 2.8) that embodies the DMs' overall consensus of the priority of the factor:

$$\mu_{\tilde{w}_i}(x) = \min_g(\mu_{\tilde{w}_i(g)}(x)) \tag{3.75}$$

However, it is possible that the FI result is an empty set, which means there is no overall consensus among the DMs.

3.6.3   Finding Out the Partial Consensus Using Partial-Consensus Fuzzy Intersection (PCFI)

When there is no overall consensus among all DMs, the partial consensus among them, i.e., the consensus among most DMs, can be sought instead.

Definition 1 (PCFI) [20] The H/G PCFI of the i-th fuzzy priority estimated by the G DMs, i.e., $\tilde{w}_i(1) \sim \tilde{w}_i(G)$ is indicated with $\tilde{I}^{H/G}(\tilde{w}_i(1), \ldots, \tilde{w}_i(G))$ such that

$$\mu_{\tilde{I}^{H/G}(\tilde{w}_i(1),\ldots,\tilde{w}_i(G))}(x) = \max_{\text{all } \upsilon}(\min(\mu_{\tilde{w}_1(\upsilon(1))}(x), \ldots, \mu_{\tilde{w}_1(\upsilon(H))}(x))) \tag{3.76}$$

where $\upsilon() \in Z^+$; $1 \le \upsilon() \le G$; $\upsilon(p) \cap \upsilon(q) = \emptyset \ \forall \ p \ne q$; $H \ge 2$.

For example, the 2/3 PCFI of $\tilde{w}_i(1) \sim \tilde{w}_i(3)$ can be obtained as

$$\mu_{\tilde{I}^{2/3}(\tilde{w}_i(1),\ldots,\tilde{w}_i(3))}(x) = \max(\min(\mu_{\tilde{w}_i(1)}(x), \mu_{\tilde{w}_i(2)}(x)), \min(\mu_{\tilde{w}_i(1)}(x), \mu_{\tilde{w}_i(3)}(x)),$$
$$\min(\mu_{\tilde{w}_i(2)}(x), \mu_{\tilde{w}_i(3)}(x))) \tag{3.77}$$

The following properties hold for the *H/G* PCFI result [22]:

- FI is equivalent to *G/G* PCFI.
- $H - 1$ membership functions are outside the *H/G* PCFI result. In contrast, $G - 1$ membership functions are outside the FI result.
- The range of $\tilde{I}^{H_1/G}(\tilde{w}_i(1), \ldots, \tilde{w}_i(G))$ is wider than that of $\tilde{I}^{H_2/G}(\tilde{w}_i(1), \ldots, \tilde{w}_i(G))$ if $H_1 < H_2$.
- The range of any PCFI result is obviously wider than that of the FI result.
- For the training data, every PCFI result contains the actual values.
- For the testing data, the probability that actual values are contained is higher in a PCFI result than in the FI result.

In addition, the PCFI aggregator meets four requirements: Boundary, monotonicity, commutativity, and associativity. The PCFI result is also a polygon-shaped fuzzy number (refer to Fig. 2.9). Compared with the original TFNs, the PCFI result has a narrower range while still containing the actual value. Therefore, the precision of estimating the fuzzy priority will be improved after applying the PCFI. In addition, it is possible to find partial consensus among the DMs using the PCFI even if there is no overall consensus when the FI is applied.

3.6.4   Defuzzifying the Aggregation Result Using the Center of Gravity (COG) Method

The PCFI result can be represented as the union of several non-normal trapezoidal fuzzy numbers (see Fig. 3.4):

$$\tilde{w}_i = \{(x_r, \mu_{\tilde{w}_i}(x_r)) | r = 1 \sim R\} \tag{3.78}$$

where $(x_r, \mu_{\tilde{w}_i}(x_r))$ is the r-th endpoint of $\tilde{w}_i$; $x_r \leq x_{r+1}$.
For any x value,

$$\mu_{\tilde{w}_i}(x) = \begin{cases} 0 & \text{if } x < x_1 \\ \mu_{\tilde{w}_i}(x_r) + \frac{x-x_r}{x_{r+1}-x_r}(\mu_{\tilde{w}_i}(x_{r+1}) - \mu_{\tilde{w}_i}(x_r)) & \text{if } x_r \leq x < x_{r+1} \\ 0 & \text{if } x_R \leq x \end{cases} \tag{3.79}$$

COG is applied to defuzzify $\tilde{w}_i$:

$$\text{COG}(\tilde{w}_i) = \frac{\int_0^1 x \cdot \mu_{\tilde{w}_i}(x)\mathrm{d}x}{\int_0^1 \mu_{\tilde{w}_i}(x)\mathrm{d}x}$$

Fig. 3.4 Representing the PCFI result with several non-normal trapezoidal fuzzy numbers

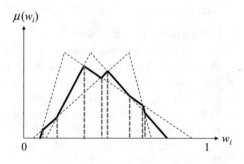

$$= \frac{\sum_{r=1}^{R} \int_{x_{r-1}}^{x_r} x \cdot \mu_{\tilde{w}_i}(x)\mathrm{d}x}{\sum_{r=1}^{R} \int_{x_{r-1}}^{x_r} \mu_{\tilde{w}_i}(x)\mathrm{d}x} \tag{3.80}$$

to which the following theorems are helpful.

Theorem 1 [22] *Let $\tilde{A}$ be a non-normal TrFN as shown in Fig. 3.4. Then the integral of $\tilde{A}$ is*

$$\int_{x_1}^{x_2} \mu_{\tilde{A}(x)}(x)\mathrm{d}x = \frac{\mu_2 x_2^2 + \mu_1 x_2^2 - 2\mu_2 x_1 x_2 + \mu_1 x_1^2 - 2\mu_1 x_1 x_2 + \mu_2 x_1^2}{2(x_2 - x_1)} \tag{3.81}$$

Theorem 2 [22] *Let $\tilde{A}$ be a non-normal TrFN as shown in Fig. 3.4. Then the integral of $x\tilde{A}$ is*

$$\int_{x_1}^{x_2} x\mu_{\tilde{A}(x)}(x)\mathrm{d}x = \frac{2\mu_2 x_2^3 + \mu_1 x_2^3 - 3\mu_2 x_1 x_2^2 + \mu_2 x_1^3 + 2\mu_1 x_1^3 - 3\mu_1 x_1^2 x_2}{6(x_2 - x_1)}$$

$$\tag{3.82}$$

References

1. R. Mamlook, O. Badran, E. Abdulhadi, A fuzzy inference model for short-term load forecasting. Energ. Policy **37**(4), 1239–1248 (2009)
2. A.K. Lohani, N.K. Goel, K.K.S. Bhatia, Improving real time flood forecasting using fuzzy inference system. J. Hydrol. **509**, 25–41 (2014)
3. P.C. Chang, C.H. Liu, A TSK type fuzzy rule based system for stock price prediction. Expert Syst. Appl. **34**(1), 135–144 (2008)
4. M. Firat, M.E. Turan, M.A. Yurdusev, Comparative analysis of fuzzy inference systems for water consumption time series prediction. J. Hydrol. **374**(3–4), 235–241 (2009)
5. A.I. Arciniegas, I.E.A. Rueda, Forecasting short-term power prices in the Ontario Electricity Market (OEM) with a fuzzy logic based inference system. Utilities Policy **16**(1), 39–48 (2008)
6. F.J. Chang, Y.T. Chang, Adaptive neuro-fuzzy inference system for prediction of water level in reservoir. Adv. Water Resour. **29**(1), 1–10 (2006)
7. M.A. Boyacioglu, D. Avci, An adaptive network-based fuzzy inference system (ANFIS) for the prediction of stock market return: the case of the Istanbul stock exchange. Expert Syst. Appl. **37**(12), 7908–7912 (2010)
8. R. Singh, A. Kainthola, T.N. Singh, Estimation of elastic constant of rocks using an ANFIS approach. Appl. Soft Comput. **12**(1), 40–45 (2012)
9. B.B. Ekici, U.T. Aksoy, Prediction of building energy needs in early stage of design by using ANFIS. Expert Syst. Appl. **38**(5), 5352–5358 (2011)
10. T. Chen, Y.C. Wang, Incorporating the FCM–BPN approach with nonlinear programming for internal due date assignment in a wafer fabrication plant. Robot. Comput.-Integr. Manuf. **26**(1), 83–91 (2010)
11. T. Chen, Forecasting the yield of a semiconductor product with a collaborative intelligence approach. Appl. Soft Comput. **13**(3), 1552–1560 (2013)

12. T. Chen, Y.C. Wang, H.R. Tsai, Lot cycle time prediction in a ramping-up semiconductor man-
    ufacturing factory with a SOM–FBPN-ensemble approach with multiple buckets and partial
    normalization. Int. J. Adv. Manuf. Technol. **42**(11–12), 1206–1216 (2009)
13. R. Fletcher, C.M. Reeves, Function minimization by conjugate gradients. Comput. J. **7**(2),
    149–154 (1964)
14. T. Chen, An effective fuzzy collaborative forecasting approach for predicting the job cycle time
    in wafer fabrication. Comput. Ind. Eng. **66**(4), 834–848 (2013)
15. T. Chen, A collaborative fuzzy-neural system for global CO_2 concentration forecasting. Int. J.
    Innov. Comput. Inf. Control **8**(11), 7679–7696 (2012)
16. P.J.M. Van Laarhoven, W. Pedrycz, A fuzzy extension of Saaty's priority theory. Fuzzy Sets
    Syst. **11**, 229–241 (1983)
17. M.A.B. Promentilla, T. Furuichi, K. Ishii, N. Tanikawa, A fuzzy analytic network process for
    multi-criteria evaluation of contaminated site remedial countermeasures. J. Environ. Manag.
    88, 479–495 (2008)
18. V. Jain, S. Sakhuja, N. Thoduka, R. Aggarwal, A.K. Sangaiah, Supplier selection using fuzzy
    AHP and TOPSIS: a case study in the Indian automotive industry. Neural Comput. Appl. **29**,
    555–564 (2016)
19. C. Kahraman, U. Cebeci, D. Ruan, Multi-attribute comparison of catering service companies
    using fuzzy AHP: The case of Turkey. Int. J. Prod. Econ. **87**, 171–184 (2004)
20. J. Ignatius, A. Hatami-Marbini, A. Rahman, L. Dhamotharan, P. Khoshnevis, A fuzzy decision
    support system for credit scoring. Neural Comput. Appl. **29**, 921–937 (2016)
21. N. Foroozesh, R. Tavakkoli-Moghaddam, S.M. Mousavi, A novel group decision model based
    on mean–variance–skewness concepts and interval-valued fuzzy sets for a selection problem
    of the sustainable warehouse location under uncertainty. Neural Comput. Appl. **30**, 3277–3293
    (2017)
22. Y.C. Wang, T. Chen, A partial-consensus posterior-aggregation FAHP method—Supplier selec-
    tion problem as an example. Mathematics **7**(2), 179 (2019)
23. T.L. Saaty, *The Analytic Hierarchy Process* (McGraw-Hill Education, New York, 1980)
24. W.C. Wedley, Consistency prediction for incomplete AHP matrices. Math. Comput. Model.
    17, 151–161 (1993)
25. I.S. Cheng, Y. Tsujimura, M. Gen, T. Tozawa, An efficient approach for large scale project
    planning based on fuzzy Delphi method. Fuzzy Sets Syst. **76**, 277–288 (1995)
26. A. Kaufmann, M.M. Gupta, *Fuzzy Mathematical Models in Engineering and Management
    Science* (North-Holland, Amsterdam, 1998)
27. T. Chen, A SOM-FBPN-ensemble approach with error feedback to adjust classification for
    wafer-lot completion time prediction. Int. J. Adv. Manuf. Technol. **37**(7–8), 782–792 (2008)
28. T. Chen, Y.C. Lin, A fuzzy-neural system incorporating unequally important expert opinions
    for semiconductor yield forecasting. Int. J. Uncertainty Fuzziness Knowl. Based Syst. **16**(01),
    35–58 (2008)
29. E. Ostrosi, J.B. Bluntzer, Z. Zhang, J. Stjepandić (2018) Car style-holon recognition in
    computer-aided design. J. Comput. Design Eng.
30. Z. Zhang, D. Xu, E. Ostrosi, L. Yu, B. Fan (2017) A systematic decision-making method for
    evaluating design alternatives of product service system based on variable precision rough set.
    J. Intell. Manuf.
31. T. Chen, Y.C. Wang, An agent-based fuzzy collaborative intelligence approach for precise and
    accurate semiconductor yield forecasting. IEEE Trans. Fuzzy Syst. **22**(1), 201–211 (2014)

Chapter 4
Fuzzy Clustering and Fuzzy Co-clustering

4.1 Introduction

Cluster analysis (or also called clustering) [1, 2] is often utilized in the first step of data analysis with the goal of summarizing structural information of data sets. Various clustering algorithms are roughly divided into *hierarchical algorithms* and *non-hierarchical algorithms*. Hierarchical algorithms are well suited for constructing classification trees such as *dendrogram* based on constructive approaches of merging distributed objects one-by-one or destructive approaches of separating clusters into subclusters. For example, constructive algorithms start from disjoint clusters of solo object each and construct clusters by merging the most similar subclusters based on such distance criteria as nearest neighbor, furthest neighbor, or other modified distances. While dendrogram provides rich structural information of data sets, hierarchical algorithms are sometimes only applicable for small data sets because of heavy computational costs in iterative construction/separation steps.

Non-hierarchical algorithms are generally designed for estimating the pre-defined number of clusters with fewer computational costs. k-Means [3] is a basic non-hierarchical algorithm, where a pre-defined number k of clusters is represented by their prototypes of *mean vectors* and the sum of within-cluster errors among within-cluster objects and prototypes are minimized. The local optimal partition is found by the Picard iteration of *nearest prototype assignment* and *mean vector calculation*. Due to the simplicity of the calculation model, k-Means is often suitable for handling large data sets rather than hierarchical algorithms. Fuzzy clustering extended the k-Means concept by introducing the *fuzziness* of cluster assignment of objects such that fuzzy memberships depict the degree of belongingness of objects to clusters. Fuzzy c-Means (FCM) [4] is a representative extension of k-Means, which has been shown to have lower initialization sensitivity than k-Means.

Recently, co-cluster structure analysis becomes popular in cooccurrence information analysis such as document–keyword analysis and gene expression analysis. Considering cooccurrence frequencies among objects and items, the goal is not only to extract clusters of familiar objects but also to characterize meaningful items in each cluster. Fuzzy clustering for categorical multivariate data (FCCM) [5] is an

FCM-type co-clustering model, which constructed a modified FCM-like objective function of the aggregation measure among objects and items. Dual fuzzy memberships of objects and items are estimated under the Picard iteration of iterating the memberships of objects and items without utilizing cluster prototypes.

In this chapter, following the brief introduction of fuzzy c-Means (FCM) clustering, FCM-induced fuzzy co-clustering model is reviewed with illustrative examples.

4.2   FCM-Type Fuzzy Clustering

4.2.1   k-Means Family

Assume that we have n objects to be partitioned into C clusters, where object i ($i = 1, \ldots, n$) is characterized by m-dimensional numeric observation $x_i = (x_{i1}, \ldots, x_{im})^\top$. k-Means [3] is a basic method for non-hierarchical clustering, where each cluster c is represented by the prototype of centroid b_c. The algorithm starts with random initialization of object C-partitions or centroid locations and iterates *nearest prototype assignment* of objects and *updating of centroids* in each cluster until convergence.

From the viewpoint of mathematical optimization, the k-Means process is reduced to minimization of the following cost function:

$$J_{km} = \sum_{c=1}^{C} \sum_{i \in G_c} ||x_i - b_c||^2, \tag{4.1}$$

where each object x_i is assigned to one of the C clusters G_c, $c = 1, \ldots, C$, and b_c is the mean vector of cluster G_c.

If we represent a C-partition of objects with membership parameter u_{ci} ($u_{ci} \in \{0, 1\}$), the cost function is rewritten as:

$$J_{km} = \sum_{c=1}^{C} \sum_{i=1}^{n} u_{ci} ||x_i - b_c||^2, \tag{4.2}$$

where u_{ci} represents the membership of object i to cluster G_c such that $u_{ci} = 1$ implies the assignment of object i to cluster c while $u_{ci} = 0$ otherwise. The exclusive assignment of objects is supported by the constraint of $\sum_{c=1}^{C} u_{ci} = 1, \forall i$.

The k-Means algorithm is a Picard iterative process for finding a local optimal solution and often derives different C-partitions from different initializations. Additionally, we can have various C-partitions with many candidates of cluster number C. So, the k-Means algorithm should be implemented with various initializations and cluster numbers, and then, the most plausible solution can be selected by considering the validity of cluster partitions with some external criteria [6].

4.2.2 Fuzzy c-Means

In fuzzy clustering [7, 8], the crisp C-partition of k-Means clustering is enhanced to a fuzzy C-partition such that fuzzy membership u_{ci} ($u_{ci} \in [0, 1]$) represents the degree of belongingness of object i to cluster c and each object can belong to multiple clusters with some degrees. Here, Eq. (4.2) is a linear function with respect to u_{ci} and the k-Means optimization problem always give a solution with extremal values $u_{ci} \in \{0, 1\}$. Then, in fuzzy c-Means (FCM) [4], the k-Means objective function is modified by introducing the nonlinearity with respect to u_{ci}.

The objective function to be minimized is defined as the membership-weighted sum of within-cluster-errors as:

$$J_{fcm} = \sum_{c=1}^{C} \sum_{i=1}^{n} u_{ci}^{\theta} ||x_i - b_c||^2, \tag{4.3}$$

where u_{ci} is given under the constraint of $\sum_{c=1}^{C} u_{ci} = 1$, $\forall i$. Because the constraint implies u_{ci} to be identified with the probability of object i drawn from cluster c among all C clusters, it is often called a *probabilistic constraint*.

θ ($\theta > 1$) is the fuzzification penalty such that a larger θ brings a fuzzier partition. The model is reduced to the conventional (crisp) k-Means with $\theta \to 1$, where $u_{ci} = 1$ for the nearest cluster. Then, the nonlinearity of the objective function J_{fcm} contributes to the membership fuzzification with $\theta > 1$. A general recommendation for θ value is $\theta = 2.0$.

The updating rules in the Picard iteration are given as follows:

$$u_{ci} = \left(\sum_{\ell=1}^{C} \left(\frac{||x_i - b_c||^2}{||x_i - b_\ell||^2} \right)^{\frac{1}{1-\theta}} \right)^{-1}, \tag{4.4}$$

$$b_c = \frac{\sum_{i=1}^{n} u_{ci}^{\theta} x_i}{\sum_{i=1}^{n} u_{ci}^{\theta}}. \tag{4.5}$$

The FCM algorithm also suffers from the initialization sensitivity and can bring different solutions from different initializations. Then, the validity of fuzzy C-partitions is evaluated with some validity measures [9].

4.2.3 FCM Variants with Other Nonlinearity Concepts

The membership fuzzification can also be achieved based on other nonlinearity concepts with regularization. The entropy-based approach [8, 10] nonlinearized the

k-Means objective function by introducing an entropy-based penalty term with an adjustable weight as:

$$J_{efcm} = \sum_{c=1}^{C} \sum_{i=1}^{n} u_{ci} ||x_i - b_c||^2 + \lambda \sum_{c=1}^{C} \sum_{i=1}^{n} u_{ci} \log u_{ci}, \qquad (4.6)$$

where λ ($\lambda > 1$) tunes the degree of fuzziness of memberships such that a larger λ brings a fuzzier partition while the model is reduced to the conventional (crisp) k-Means with $\lambda \to 0$.

The updating rules in the Picard iteration are given as follows:

$$u_{ci} = \frac{\exp\left(-\frac{1}{\lambda}||x_i - b_c||^2\right)}{\sum_{\ell=1}^{C} \exp\left(-\frac{1}{\lambda}||x_i - b_\ell||^2\right)}, \qquad (4.7)$$

$$b_c = \frac{\sum_{i=1}^{n} u_{ci} x_i}{\sum_{i=1}^{n} u_{ci}}. \qquad (4.8)$$

This model is reduced to a type of Gaussian mixture models (GMMs) [11, 12] when λ equals to the double of the cluster-wise variances. In this sense, the value of λ can be recommended to be close to the double of the cluster-wise variances.

Besides, the close connection between FCM clustering and statistic models can contribute to the improvement of the conventional statistical models, e.g., supported by the deterministic annealing process [13], robust estimation of GMMs can be achieved. We can also consider a fuzzy counterpart of GMMs with full parameters: mean vectors, covariance matrices, and mixture weights, which are updated in a similar concept to the expectation–maximization (EM) algorithm [14].

Similar fuzzification models have also been proposed utilizing such nonlinear terms as a quadric penalty [15] and a K-L information-based penalty [16, 17].

4.2.4 FCM Variants with Non-point Prototypes

FCM clustering can be also extended in different directions by considering other types of prototypes rather than the point prototype of FCM. If we replace prototypical centroids of FCM with lines or hyper-planes, FCM is extended to fuzzy c-Lines (FCL) and fuzzy c-Varieties (FCV) [4], where the FCM process is modified by replacing prototype estimation with intra-cluster subspace learning [17]. Especially, FCL and

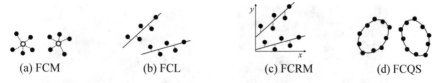

(a) FCM   (b) FCL   (c) FCRM   (d) FCQS

Fig. 4.1 Comparison of basic FCM and some FCM variants with non-point prototypes

FCV are identified with local principal component analysis (local PCA) [18, 19], which simultaneously performs *data partitioning* and *PCA in local area*.

Fuzzy c-Regression Models (FCRM) [20] is an FCM-type counterpart of switching regression model [21], where prototypical centroids of FCM is replaced with *local regression models* and regression errors are adopted not only for regression model estimation but also for clustering criterion.

For image processing, we can utilize various shape prototypes such as quadric shells (Fuzzy c-Quadric Shells: FCQS) and rectangular shells (Fuzzy c-Rectangular Shells: FCRS) [7]. For other variants, see related references (c.f., [22]) (Fig. 4.1).

4.2.5 Examples of FCM Implementation

Assume that we have a two-dimensional data set composed of three Gaussian distributions with 50 objects each as shown in Fig. 4.2. The three Gaussian subsets have mean vectors of $(0.0, 8.7)$, $(5.0, 0.0)$ and $(-5.0, 0.0)$ with spherical variances of $\begin{pmatrix} 1.0 & 0.0 \\ 0.0 & 1.0 \end{pmatrix}$, $\begin{pmatrix} 4.0 & 0.0 \\ 0.0 & 4.0 \end{pmatrix}$ and $\begin{pmatrix} 4.0 & 0.0 \\ 0.0 & 4.0 \end{pmatrix}$, respectively.

Figure 4.3 compares fuzzy C-partitions given by the standard fuzzy c-Means with different fuzzification degrees, where grayscale color represents the maximum membership values $u_{\max} = \max\{u_1, u_2, u_3\}$ at each location. Each black diamond depicts

Fig. 4.2 A two-dimensional sample distribution

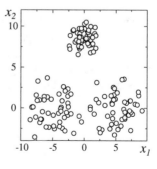

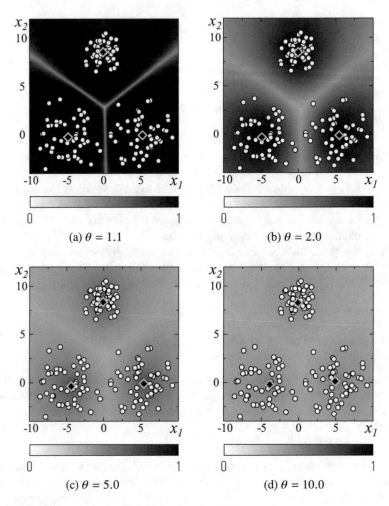

Fig. 4.3 Comparison of fuzzy C-partitions given by FCM

the prototypical centroids of clusters. When fuzzification penalty θ is close to 1 such as $\theta = 1.1$, the fuzzy C-partition is almost the nearest neighbor assignment of (crisp) k-Means, where the data partition is associated with a Voronoi tessellation of the data space. As the fuzziness degree becomes larger, the boundaries of clusters become *fuzzy* and almost all objects are equally shared by all clusters with very large θ.

4.3 FCM-Type Fuzzy Co-clustering

4.3.1 Cooccurrence Information Analysis

Recently, cooccurrence information analysis becomes popular in many fields such as document analysis, market analysis, and genomic analysis, where the goal is to reveal the intrinsic connection among two sources of *objects* and *items*. For example, in document analysis, we often have the bag-of-words information [23] of word (*item*) frequencies in each document (*object*). User-product purchase history in market analysis and a type of gene expression data are also stored in similar forms.

Assume that we have a cooccurrence information matrix of Table 4.1, where 1 implies such information as appearance of keyword j in document i and 0 otherwise. If we arrange the matrix as Table 4.2, we can find two object groups of $\{a, b, e\}$ and $\{c, d, f\}$ such that objects $\{a, b, e\}$ are connected through items $\{1, 5, 6, 2\}$ while objects $\{c, d, f\}$ are connected through items $\{2, 3, 4\}$. This type of cooccurrence information is utilized such tasks in document summarization [24] and collaborative filtering [25].

Co-clustering aims to extract co-clusters composed of mutually familiar objects and items such that the cooccurrence degrees of objects and items are high within a co-cluster. Then, the goal of co-clustering is to simultaneously estimate the cluster assignment of both objects and items such that the degree of aggregation of objects and items are high within a co-cluster while not so high inter-co-clusters.

Table 4.1 A sample of cooccurrence information matrix

Item		1	2	3	4	5	6
Object	a	1	1	0	0	0	1
	b	1	1	0	0	1	0
	c	0	1	1	0	0	0
	d	0	1	1	1	0	0
	e	0	0	0	0	1	1
	f	0	1	0	1	0	0

Table 4.2 Arranged matrices of Table 4.1

Item		1	5	6	2	3	4
Object	a	1	0	1	1	0	0
	b	1	1	0	1	0	0
	e	0	1	1	0	0	0
	c	0	0	0	1	1	0
	d	0	0	0	1	1	1
	f	0	0	0	1	0	1

4.3.2   FCCM and Fuzzy CoDoK

Assume that we have cooccurrence information of n objects with m items and is summarized in an $n \times m$ matrix $R = \{r_{ij}\}$, where r_{ij} represents the cooccurrence degree among object i and item j. In order to represent the dual partition of objects and items, FCM-type fuzzy co-clustering considers two types of fuzzy memberships: u_{ci} for the membership of object i to cluster c and w_{cj} for the membership of item j to cluster c.

Fuzzy clustering for categorical multivariate data (FCCM) [5] measures the degree of aggregation of objects and items in cluster c as:

$$Aggregation_c = \sum_{i=1}^{n} \sum_{j=1}^{m} u_{ci} w_{cj} r_{ij}, \qquad (4.9)$$

which becomes larger when both object i and item j having large cooccurrence r_{ij} have large memberships u_{ci} and w_{cj} in a certain cluster c.

Here, we must note that the sum of aggregation $\sum_c Aggregation_c$ has a trivial maximum $\sum_{i=1}^{n} \sum_{j=1}^{m} r_{ij}$ if we consider the same probabilistic constraints for both objects and items such that $\sum_{c=1}^{C} u_{ci} = 1, \forall i$ and $\sum_{c=1}^{C} w_{cj} = 1, \forall j$. That is, $\sum_c Aggregation_c$ has the maximum value when all objects and items are gathered into a certain cluster.

In order to avoid such *whole one cluster*, two types of fuzzy memberships generally have different meanings under different constraints. u_{ci} is estimated under the same constraint $\sum_{c=1}^{C} u_{ci} = 1, \forall i$ with FCM such that u_{ci} can be identified with the probability of belongingness of object i to cluster c among C clusters. On the other hand, w_{cj} is estimated under a different type constraint $\sum_{j=1}^{m} w_{cj} = 1, \forall c$ such that w_{cj} represents the typicality of item j in cluster c. Then, w_{cj} mainly contributes to the characterization of each cluster like prototypical centroids in FCM. The concept is also similar to such statistical co-clustering models as multinomial mixture models (MMMs) [26] and its variants [27].

FCCM defined the following objective function to be maximized:

$$J_{fccm} = \sum_{c=1}^{C} \sum_{i=1}^{n} \sum_{j=1}^{m} u_{ci} w_{cj} r_{ij} - \lambda_u \sum_{c=1}^{C} \sum_{i=1}^{n} u_{ci} \log u_{ci} - \lambda_w \sum_{c=1}^{C} \sum_{j=1}^{m} w_{cj} \log w_{cj},$$

$$(4.10)$$

where the entropy-based penalty terms nonlinearize the aggregation measure for dual membership fuzzification in the same manner with the entropy-based FCM [10]. λ_u and λ_w are the penalty weights for tuning the degrees of fuzziness of object and item partitions, respectively.

The updating rules in the Picard iteration are given as follows:

$$u_{ci} = \frac{\exp\left(\frac{1}{\lambda_u} \sum_{j=1}^{m} w_{cj} r_{ij}\right)}{\sum_{\ell=1}^{C} \exp\left(\frac{1}{\lambda_u} \sum_{j=1}^{m} w_{\ell j} r_{ij}\right)}, \tag{4.11}$$

$$w_{cj} = \frac{\exp\left(\frac{1}{\lambda_w} \sum_{i=1}^{n} u_{ci} r_{ij}\right)}{\sum_{\ell=1}^{m} \exp\left(\frac{1}{\lambda_w} \sum_{i=1}^{n} u_{ci} r_{i\ell}\right)}. \tag{4.12}$$

Fuzzy co-clustering of documents and keywords (Fuzzy CoDoK) [28] is a modification of FCCM, where linear aggregation measure $\sum_c Aggregation_c$ is nonlinearized by a quadric penalty [15]. Although this method is useful for handling large scale data sets by improving computational stability of Eqs.(4.11) and (4.12), their updating formulas need some tricks for avoiding negative fuzzy memberships. Other fuzzification methods are also available using such nonlinear penalties as Kullback–Leibler (KL) divergence [29] and q-divergence [30].

The standard FCM-type fuzzification is also possible in FCCM, where the objective function to be maximized is nonlinearized with the fuzzification exponent θ ($\theta < 1$) [31].

4.3.3 Fuzzy Co-clustering with Statistical Concepts

Other fuzzification schemes have also been proposed by considering different non-linearization strategies. Fuzzy co-clustering induced by multinomial mixture models (FCCMM) [32] adopted a modified aggregation measure based on the MMMs concept [26] and realized a fuzzy counterpart of statistical MMMs.

Assume that we have an $n \times m$ cooccurrence matrix $R = \{r_{ij}\}$, where n objects were drawn from one of C multinomial distributions and the probability of occurrence of item j in model c is given as w_{cj} such that $\sum_{j=1}^{m} w_{cj} = 1$. If object i is drawn from model c with probability u_{ci} such that $\sum_{c=1}^{C} u_{ci} = 1$, maximum likelihood estimation for u_{ci} and w_{cj} is derived from maximization of the following pseudo-log-likelihood:

$$J_{mmms} = \sum_{c=1}^{C} \sum_{i=1}^{n} \sum_{j=1}^{m} u_{ci} r_{ij} \log w_{cj} + \sum_{c=1}^{C} \sum_{i=1}^{n} u_{ci} \log \frac{\alpha_c}{u_{ci}}, \tag{4.13}$$

where α_c is the a priori probability of model c. Based on the EM iteration [14], the maximum likelihood estimators for u_{ci}, α_c and w_{cj} are calculated.

Although the pseudo-log-likelihood function was constructed based purely on the probabilistic concept, from the fuzzy clustering viewpoint, the first term is almost identified with the aggregation measure of Eq. (4.9) except for the log function and the second K-L information term works for nonlinearization of the aggregation term with respect to u_{ci}, i.e., fuzzification of u_{ci}. Then, we can expect to achieve various fuzzy partitions by changing the responsibility of the second penalty term. Introducing an adaptable penalty weight, the objective function of FCCMM was proposed as follows:

$$J_{fccmm} = \sum_{c=1}^{C} \sum_{i=1}^{n} \sum_{j=1}^{m} u_{ci} r_{ij} \log w_{cj} + \lambda_u \sum_{c=1}^{C} \sum_{i=1}^{n} u_{ci} \log \frac{\alpha_c}{u_{ci}}, \qquad (4.14)$$

where penalty weight λ_u tunes the degree of fuzziness of object partition such that a larger λ_u ($\lambda_u > 1$) implies a fuzzier object partition while a smaller λ_u ($\lambda_u < 1$) brings a k-Means-like crisp partition.

Additionally, the fuzziness degree of item partition can also be tuned by changing the degree of nonlinearity with respect to w_{cj}. Considering the definition of log function:

$$\log w_{cj} = \lim_{\lambda_w \to 0} \frac{1}{\lambda_w} \left((w_{cj})^{\lambda_w} - 1 \right), \qquad (4.15)$$

we can tune the fuzziness degree of item partition by changing λ_w of the right-hand side of Eq. (4.15) such that a smaller λ_w ($\lambda_w < 0$) implies a fuzzier item partition while $\lambda_w \to 1$ brings k-Means-like crisp partition; i.e., each co-cluster is characterized only by a solo item.

Similar models for tuning the intrinsic fuzziness of statistical co-clustering model can be also achieved in other mixture models such as probabilistic latent semantics (pLSA) [33]. These close connection between fuzzy co-clustering and statistic models can also contribute to the improvement of statistical co-clustering model estimation [34].

4.3.4   Examples of FCCM Implementation

Assume that we have a cooccurrence information matrix $R = \{r_{ij}\}$ among 100 objects and 70 items as shown in Fig. 4.4, where black and white cells mean $r_{ij} = 1$ and $r_{ij} = 0$, respectively. In this data set, we can find roughly four co-clusters in the diagonal positions, which are somewhat sparse and are varied in noise. All objects

Fig. 4.4 A sample
cooccurrence information
matrix

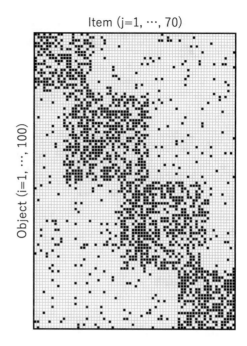

are drawn from one of the four co-clusters while some items are related to multiple co-clusters. This kind of situation is often found in such tasks as document clustering, where each document is related to a solo topic while some general keywords are typical in multiple topics. The goal is to extract the co-clusters by simultaneously estimating fuzzy memberships of both objects u_{ci} and items w_{cj}.

Figure 4.5 shows the two types of fuzzy memberships given by the FCCM algorithm with $\lambda_u = 0.1$ and $\lambda_w = 1.0$. In the figure, grayscale color depicts the degree of fuzzy memberships such as black for the maximum value and white for zero memberships. Figure 4.5a indicates that 100 objects were successfully partitioned into four visual clusters while some boundary objects were weakly shared by multiple clusters. Due to the sum-to-one condition for C clusters, object memberships are available for capturing object assignment to clusters.

On the other hand, item memberships shown in Fig. 4.5b are designed for evaluating the typicality of each item in co-clusters under the sum-to-one condition with respect to items. This type of fuzzy memberships is useful for finding cluster-wise typical items in each cluster and Fig. 4.5b successfully reveals both intra-cluster typicality of each item and inter-cluster sharing of some items.

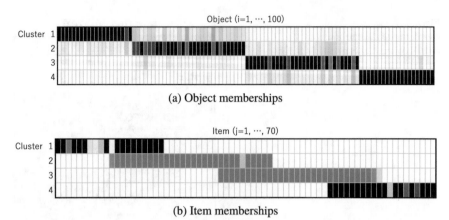

(a) Object memberships

(b) Item memberships

Fig. 4.5 Two types of fuzzy memberships given by FCCM

4.4   Summary

This chapter presented a brief review of FCM-type clustering and its extension to fuzzy co-clustering in conjunction with some illustrative examples. Fuzzy co-clustering is becoming a more important tool for cooccurrence information analysis in Web data mining.

In this chapter, the close relation among FCM-type fuzzy clustering and some statistical models was also reviewed. By carefully tuning the intrinsic fuzziness degree, FCM-type fuzzy clustering can also contribute to the robust estimation of statistical models.

In the next chapter, some collaborative models for performing FCM-type clustering are introduced, where distributed databases are jointly handled among multiple organizations.

References

1. M.R. Anderberg, *Cluster Analysis for Applications* (Academic Press, New York, 1973)
2. A.K. Jain, R.C. Dubes, *Algorithms for Clustering Data* (Prentice Hall, Englewood Cliffs, NJ, 1988)
3. J.B. MacQueen, Some methods of classification and analysis of multivariate observations, in *Proceeding of 5th Berkeley Symposium on Mathematics of Stats and Probability* (1967), pp. 281–297
4. J.C. Bezdek, *Pattern Recognition with Fuzzy Objective Function Algorithms* (Plenum Press, 1981)
5. C.-H. Oh, K. Honda, H. Ichihashi: Fuzzy clustering for categorical multivariate data, in *Proceeding of Joint 9th IFSA World Congress and 20th NAFIPS International Conference* (2001), pp. 2154–2159
6. J.C. Dunn, Well separated clusters and optimal fuzzy partitions. J. Cybern. **4**, 95–104 (1974)

7. F. Höppner, F. Klawonn, R. Kruse, T. Runkler, *Fuzzy Cluster Analysis* (Wiley, 1999)
8. S. Miyamoto, H. Ichihashi, K. Honda, *Algorithms for Fuzzy Clustering* (Springer, 2008)
9. W. Wang, Y. Zhang, On fuzzy cluster validity indices. Fuzzy Sets Syst. **158**, 2095–2117 (2007)
10. S. Miyamoto, M. Mukaidono, Fuzzy c-means as a regularization and maximum entropy approach, in *Proceeding of the 7th International Fuzzy Systems Association World Congress*, vol. 2. (1997), pp. 86–92
11. R.J. Hathaway, Another interpretation of the EM algorithm for mixture distributions. Stat. & Probab. Lett. **4**, 53–56 (1986)
12. C.M. Bishop, *Neural Networks for Pattern Recognition* (Clarendon Press, 1995)
13. K. Rose, E. Gurewitz, G. Fox, A deterministic annealing approach to clustering. Pattern Recognit. Lett. **11**, 589–594 (1990)
14. A.P. Dempster, N.M. Laird, D.B. Rubin, Maximum likelihood from incomplete data via the EM algorithm, J. R. Stat. Soc., Series B, **39**, 1–38 (1977)
15. S. Miyamoto, K. Umayahara, Fuzzy clustering by quadratic regularization, in *Proceeding 1998 IEEE International Conference Fuzzy Systems and IEEE World Congress. Computational Intelligence*, vol. 2. (1998) pp. 1394–1399
16. H. Ichihashi, K. Miyagishi, K. Honda, Fuzzy c-means clustering with regularization by K-L information, in *Proceeding of 10th IEEE International Conference on Fuzzy Systems*, vol. 2. (2001) pp. 924–927
17. K. Honda, H. Ichihashi, Regularized linear fuzzy clustering and probabilistic PCA mixture models. IEEE Trans. Fuzzy Systems **13**(4), 508–516 (2005)
18. M.E. Tipping, C.M. Bishop, Mixtures of probabilistic principal component analysers. Neural Comput. **11**(2), 443–482 (1999)
19. K. Honda, H. Ichihashi, Linear fuzzy clustering techniques with missing values and their application to local principal component analysis. IEEE Trans. Fuzzy Systems **12**(2), 183–193 (2004)
20. R.J. Hathaway, J.C. Bezdek, Switching regression models and fuzzy clustering. IEEE Trans. on Fuzzy Systems **1**(3), 195–204 (1993)
21. R.E. Quandt, A new approach to estimating switching regressions. J. Am. Stat. Assoc. **67**, 306–310 (1972)
22. J.C. Bezdek, J. Keller, R. Krishnapuram, N.R. Pal, *Fuzzy Models and Algorithms for Pattern Recognition and Image Processing* (Kluwer, Boston, 1999)
23. G. Salton, C. Buckley, Term-weighting approaches in automatic text retrieval. Inf. Process. Manage. **24**(5), 513–523 (1988)
24. V. Gupta, G.S. Lehal, A survey of text summarization extractive techniques. J. Emerg. Technol. Web Intell. **2**(3), 258–268 (2010)
25. K. Honda, A. Notsu, H. Ichihashi, Collaborative filtering by sequential user-item co-cluster extraction from rectangular relational data. Int. J. Knowl. Eng. Soft Data Parad. **2**(4), 312–327 (2010)
26. L. Rigouste, O. Cappé, F. Yvon, Inference and evaluation of the multinomial mixture model for text clustering. Inf. Process. Manag. **43**(5), 1260–1280 (2007)
27. K. Sjölander, K. Karplus, M. Brown, R. Hughey, A. Krogh, I. Saira Mian, D. Haussler, Dirichlet mixtures: a method for improved detection of weak but significant protein sequence homology. Comput. Appl. Biosci. **12**(4), 327–345 (1996)
28. K. Kummamuru, A. Dhawale, R. Krishnapuram: Fuzzy co-clustering of documents and keywords, in *Proceeding 2003 IEEE International Conference Fuzzy Systems*, vol. 2. (2003), pp. 772–777
29. K. Honda, S. Oshio, A. Notsu, FCM-type fuzzy co-clustering by K-L information regularization, in *Proceeding of 2014 IEEE International Conference on Fuzzy Systems* (2014), pp. 2505–2510
30. T. Kondo, Y. Kanzawa, Fuzzy clustering methods for categorical multivariate data based on q-divergence. J. Adv. Comput. Intell. Intell. Inform. **22**(4), 524–536 (2018)
31. Y. Kanzawa, Bezdek-type fuzzified co-clustering algorithm. J. Adv. Comput. Intell. Intell. Inform. **19**(6), 852–860 (2015)

32. K. Honda, S. Oshio, A. Notsu, Fuzzy co-clustering induced by multinomial mixture models. J. Adv. Comput. Intell. Intell. Inform. **19**(6), 717–726 (2015)
33. T. Hofmann, Unsupervised learning by probabilistic latent semantic analysis. Mach. Learn. **42**(1–2), 177–196 (2001)
34. T. Goshima, K. Honda, S. Ubukata, A. Notsu, Deterministic annealing process for pLSA-induced fuzzy co-clustering and cluster splitting characteristics. Int. J. Approx. Reason. **95**, 185–193 (2018)

Chapter 5
Collaborative Framework for Fuzzy Co-clustering

5.1 Introduction

In cases of collaborative data analysis among multiple organizations, we must take care of personal privacy and keep secret the original raw data within each organization. But, from the practical viewpoint, we can expect to discover more reliable knowledge through collaborative use of multiple data sources.

Privacy preserving data mining (PPDM) [1] is a fundamental approach for utilizing multiple databases including personal or sensitive information without fear of information leaks. Although it is possible to modify the original databases through some anonymization approaches such as a priori k-anonymization of databases for secure publication [2, 3], such preprocessing can bring severe information losses. Another direction of utilizing all distributed information is to analyze them by sharing only the derived knowledge without revealing each element itself. In k-Means clustering [4], several secure procedures for estimating cluster prototypes have been proposed [5, 6].

In this chapter, after a brief review on k-Means-type clustering processes, collaborative frameworks for utilizing distributed cooccurrence matrices in fuzzy co-cluster structure estimation are introduced [7], where cooccurrence information among objects and items are separately stored in several sites. Cooccurrence information can be distributed through several sites in two different forms. In vertically distributed databases, all sites share common objects but they are characterized with different (independent) items in each site. On the other hand, in horizontally distributed databases, all sites share common items but each site gathers cooccurrence information on different (independent) objects. In both situations, the goal is to reveal the global co-cluster structures varied in whole separate databases without publishing each element of independent databases to other sites.

© The Author(s), under exclusive license to Springer Nature Switzerland AG 2020
T.-C. T. Chen and K. Honda, *Fuzzy Collaborative Forecasting and Clustering*,
SpringerBriefs in Applied Sciences and Technology,
https://doi.org/10.1007/978-3-030-22574-2_5

5.2 Collaborative Framework for k-Means-Type Clustering Process

Assume that we have n objects to be partitioned into C clusters, where object i ($i = 1, \ldots, n$) is characterized by m-dimensional numeric observation $x_i = (x_{i1}, \ldots, x_{im})^\top$. In k-Means clustering [4], each cluster is represented by prototypical centroid b_c, $c = 1, \ldots, C$, and the objective function to be minimized is given as the following within-group sum of errors:

$$ J_{km} = \sum_{c=1}^{C} \sum_{i=1}^{n} u_{ci} \|x_i - b_c\|^2, \tag{5.1} $$

where u_{ci} represents the membership of object i to cluster G_c such that $u_{ci} \in \{0, 1\}$, $\forall c, i$ under the probabilistic constraint $\sum_{c=1}^{C} u_{ci} = 1$, $\forall i$. Starting with random initialization, a Picard iterative process of *nearest prototype assignment of objects* and *cluster center estimation* is repeated until convergence.

Now, let's consider that the whole data information is exclusively distributed to T sites and each site cannot broadcast their data to other sites due to such reasons as privacy or patent issues. Here, we can have two different forms of distributed databases: *vertically distributed databases* and *horizontally distributed databases* (Fig. 5.1).

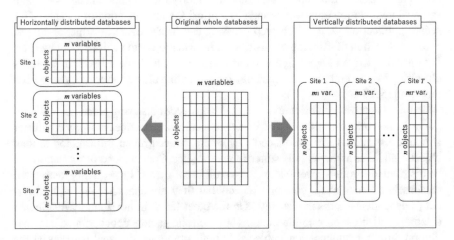

Fig. 5.1 Two different forms of distributed databases

5.2.1 Vertically Distributed Databases

In vertically distributed databases, T sites ($t = 1, \ldots, T$) have m_t-dimensional observations $x_i^t = (x_{i1}^t, \ldots, x_{im_t}^t)^\top$ on common n objects ($i = 1, \ldots, n$) and $\sum_{t=1}^T m_t = m$. If we can gather all data pieces into a whole data vector x_i, the nearest centroid in k-Means process is searched for by calculating the distances between object x_i and centroids b_c as:

$$||x_i - b_c||^2 = \sum_{t=1}^T d_{ci}^t = \sum_{t=1}^T ||x_i^t - b_c^t||^2, \qquad (5.2)$$

where $d_{ci}^t = ||x_i^t - b_c^t||^2$ is the within-cluster square distance in site t. This implies that we can find the nearest centroid only with d_{ci}^t without broadcasting each element x_{ij}^t. Then, in privacy preserving k-Means clustering, d_{ci}^t instead of x_{ij}^t are utilized for performing object assignment in a secure manner.

Now, assume the following conditions in collaborative analysis among T sites:

- All sites are willing to share the cluster assignment of all objects, i.e., u_{ci}.
- Each site cannot open their intra-site information, i.e., m_t, x_i^t, b_c^t, and d_{ci}^t.

It is obvious that each element of site-wise cluster center b_{cj}^t is securely calculated in each cluster as:

$$b_{cj}^t = \frac{\sum_{i=1}^n u_{ci} x_{ij}^t}{\sum_{i=1}^n u_{ci}}. \qquad (5.3)$$

Then, a secure process must be considered only in the object assignment step.

Vaidya and Clifton [8] adopted a simple encryption operation in calculation of object-wise distances Eq. (5.2). Assume that we have at least three non-colluding sites ($T > 2$) and they want to securely calculate the distance between object i and cluster c, i.e., the sum of $d_{ci}^t, t = 1, \ldots, T$. In this operation, site 1 and site T have their special roles. First, site 1 generates T random values $v_t, t = 1, \ldots, T$ such that $\sum_{t=1}^T v_t = 0$ and sends each v_t to site t. Second, in each site t, a distorted value $(d_{ci}^t + v_t)$ is sent to site T. Finally, site T calculates $\sum_{t=1}^T (d_{ci}^t + v_t) = \sum_{t=1}^T d_{ci}^t = ||x_i - b_c||^2$. This process can be iterated C times, and then, site T can find the nearest cluster of object x_i and broadcast u_{ci} to all sites. Due to encryption by random values v_t, the intra-site information is kept secret within each site.

By the way, in the above model, site T has a special information on *distances between objects and clusters* in conjunction with *object assignment*. If we want to avoid such a special situation, nearest prototype assignment can also be performed with secure comparison [8].

In more secure cases, much more secure process can be achieved by such techniques as *secure dot product* and *homomorphic encryption* [9]. More detailed discussion on privacy preserving clustering can be found in [6, 10]. Additionally, more general references on secure computation can be found in [11, 12].

5.2.2   Horizontally Distributed Databases

Next, in horizontally distributed databases, T sites ($t = 1, \ldots, T$) have different (independent) n_t objects with m-dimensional observations $x_i^t = (x_{i1}^t, \ldots, x_{im}^t)^\top$ on common m attributes ($i = 1, \ldots, n_t$ in site t) and $\sum_{t=1}^{T} n_t = n$. The goal of k-Means clustering is to estimate cluster centroids b_c, $c = 1, \ldots, C$ and object partition u_{ci}^t for object i of site t.

Now, assume the following conditions in collaborative analysis among T sites:

- All sites are willing to share the cluster prototypes of all clusters, i.e., b_c.
- Each site cannot open their intra-site information, i.e., n_t, x_i^t, and u_{ci}^t.

It is obvious that object assignment u_{ci}^t is securely estimated by calculating the distances between objects and cluster centroids in each cluster as:

$$||x_i^t - b_c||^2 = \sum_{j=1}^{m} (x_{ij}^t - b_{cj})^2. \qquad (5.4)$$

Then, a secure process must be considered only in the centroid estimation step.

If we can gather all data pieces into a whole data set, the cluster centroid in k-Means process is calculated as:

$$b_c = \frac{\sum_{t=1}^{T} \alpha_c^t}{\sum_{t=1}^{T} \beta_c^t} = \frac{\sum_{t=1}^{T} \sum_{i=1}^{n_t} u_{ci}^t x_i^t}{\sum_{t=1}^{T} \sum_{i=1}^{n_t} u_{ci}^t}, \qquad (5.5)$$

where $\alpha_c^t = \sum_{i=1}^{n_t} u_{ci}^t x_i^t$ and $\beta_c^t = \sum_{i=1}^{n_t} u_{ci}^t$ are the intra-site sum of site-wise information. Then, in privacy preserving k-Means clustering, cluster centroids must be calculated by utilizing α_c^t and β_c^t in a secure manner.

Besides secure summation calculation [5] as in the case of vertically distributed databases, third-party calculation model was also proposed for this type of clustering [13].

By the way, in collaborative work on clustering, handling of horizontally distributed databases can contribute to efficient clustering of large data sets. When we handle very large data sets in fuzzy c-Means (FCM) clustering [14], it is quite difficult to load all observations of a number of objects at a time. Then, several efficient approaches have been proposed [15].

The simplest approach, which is called *random sample* and *extend* approach (rse-FCM), applies the conventional FCM algorithm only to a subset of the data set [16]. However, the clustering quality might be significantly degraded through data rejection. Then, we should efficiently apply the FCM algorithm to all subsets of the whole data set.

The *single pass* approach (spFCM) [17] sequentially performs the weighted FCM algorithm to the subsets such that the cluster centers of the previous subprocess, i.e., the previous subset, are added to the next subset with a certain weights in each subprocess for inheriting the current cluster structures to the next subprocess. In cases of horizontally distributed databases among T sites, each site can sequentially perform the weighted k-Means or FCM. Let's consider a sequential k-Means process from site 1 to site T, where each site t estimates its own cluster centroids b_c^t, $c = 1, \ldots, C$. First, site 1 estimates centroids b_c^1 and their object cardinality n_c^1, and then, sends them to site 2. In site t ($t > 1$), the previous centroids are added to the site-wise dataset with weights n_c^{t-1} and k-Means is performed considering the data weights. Then, the derived centroids b_c^t are sent to site $t + 1$ with their object cardinality n_c^t, where $\sum_{c=1}^{C} n_c^t = \sum_{\ell=1}^{t} n_\ell$. The final result is given in site T, where cluster centroids are estimated from all n objects.

Besides above sampling approaches, the *online* approach (oFCM) [18] is performed in a collaborative way based on the concept of horizontally distributed databases. oFCM performs the conventional FCM with each of all subsets in parallel and re-performs the weighted FCM algorithm with the derived cluster centers in conjunction with their membership weights. In the above horizontally distributed situation, T sites can independently and parallelly perform k-Means or FCM for estimating their intra-site centroids b_c^t, $c = 1, \ldots, C$. Then, the weighted k-Means-type clustering can be implemented with all intra-site centroids considering their cardinalities. When we adopt secure calculation in the weighted clustering algorithms, the process is reduced to the above privacy preserving schemes.

5.3 Collaborative Framework for FCCM of Vertically Partitioned Cooccurrence Information

Here, the collaborative framework for k-Means is adopted in the fuzzy co-clustering context. Assume that we have cooccurrence information of n objects with m items and it is summarized in an $n \times m$ matrix $R = \{r_{ij}\}$, where r_{ij} represents the cooccurrence degree among object i and item j. In order to represent the dual partition of objects and items, FCM-type fuzzy co-clustering considers two types of fuzzy memberships: u_{ci} for the membership of object i to cluster c and w_{cj} for the membership of item j to cluster c. Fuzzy clustering for categorical multivariate data (FCCM) [19] defined the objective function considering the degree of aggregation of objects and items in each cluster and the entropy-based fuzzification as:

$$J_{fccm} = \sum_{c=1}^{C}\sum_{i=1}^{n}\sum_{j=1}^{m} u_{ci}w_{cj}r_{ij} - \lambda_u \sum_{c=1}^{C}\sum_{i=1}^{n} u_{ci}\log u_{ci} - \lambda_w \sum_{c=1}^{C}\sum_{j=1}^{m} w_{cj}\log w_{cj},$$

$$(5.6)$$

where the entropy-based penalty terms work for membership fuzzification in the same manner with the entropy-based FCM [20, 21]. Under the alternative optimization scheme, u_{ci} and w_{cj} are iteratively updated until convergence.

In the following, a secure calculation model for the FCCM clustering with vertically partitioned cooccurrence information databases is considered [7]. Assume that T sites ($t = 1, \ldots, T$) share common n objects ($i = 1, \ldots, n$) and have different cooccurrence information on different items, which are summarized into $n \times m_t$ matrices $R_t = \{r_{ij}^t\}$, where m_t is the number of items in site t and $\sum_{t=1}^{T} m_t = m$.

The goal of the task is to estimate object and item memberships as similar to the full-data case as possible by sharing object partition information without broadcasting cooccurrence information r_{ij}^t. Because object memberships u_{ci} are expected to be shared and be common in all sites, we can adopt the same condition with the conventional FCCM as $\sum_{c=1}^{C} u_{ci} = 1$, $\forall i$. On the other hand, item memberships w_{cj} are somewhat different from the conventional FCCM because they must be kept secret in each site and cannot be opened for other sites. In this section, it is assumed that item memberships w_{cj}^t are independently estimated in each site t following the site-wise constraint $\sum_{j=1}^{m_t} w_{cj}^t = 1$, where w_{cj}^t is the item membership on item j in site t.

Then, the objective function to be maximized is modified as

$$J_{fccm-vd} = \sum_{c=1}^{C}\sum_{i=1}^{n}\sum_{t=1}^{T}\sum_{j=1}^{m_t} u_{ci}w_{cj}^t r_{ij}^t - \lambda_u \sum_{c=1}^{C}\sum_{i=1}^{n} u_{ci}\log u_{ci}$$

$$-\lambda_w \sum_{c=1}^{C}\sum_{t=1}^{T}\sum_{j=1}^{m_t} w_{cj}^t \log w_{cj}^t. \qquad (5.7)$$

Considering the optimality under the sum-to-one condition, the updating formula of object membership u_{ci} is given as:

$$u_{ci} = \frac{\exp\left(\dfrac{1}{\lambda_u}\sum_{t=1}^{T}\sum_{j=1}^{m_t} w_{cj}^t r_{ij}^t\right)}{\sum_{\ell=1}^{C}\exp\left(\dfrac{1}{\lambda_u}\sum_{t=1}^{T}\sum_{j=1}^{m_t} w_{\ell j}^t r_{ij}^t\right)}. \qquad (5.8)$$

This formula implies that each object membership function is dependent on $\sum_{t=1}^{T}\sum_{j=1}^{m_t} w_{cj}^t r_{ij}^t$, which is the sum of site-wise independent information $\sum_{j=1}^{m_t} w_{cj}^t r_{ij}^t$. However, a site can know the co-cluster structures of the other sites if $\sum_{j=1}^{m_t} w_{cj}^t r_{ij}^t$ is

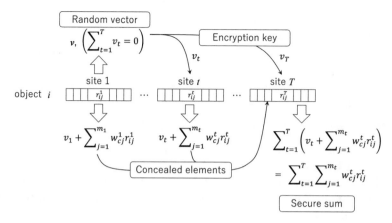

Fig. 5.2 A secure form for sum calculation in collaborative FCCM with vertically distributed databases

broadcasted without concealing their actual values, and secure calculation must be performed based on encryption operation as in the case of k-Means clustering.

For example, the simple encryption operation adopted by Vaidya and Clifton [8] can be utilized as shown in Fig. 5.2. Assume that we have at least three sites, i.e., $T > 2$, and two sites of site 1 and site T are selected as representative sites.

- Site 1 generates a length C random vector $v_t = (v_{t1}, \ldots, v_{tC})^{\top}$ for each site t, such that $\sum_{t=1}^{T} v_t = 0$.
- Sites $1 \ldots T - 1$ send $v_{tc} + \sum_{j=1}^{m_t} w_{cj}^t r_{ij}^t$ to site T.
- Their total amount $\sum_{t=1}^{T} (v_{tc} + \sum_{j=1}^{m_t} w_{cj}^t r_{ij}^t)$ is calculated for estimating u_{ci} in site T.

$\sum_{t=1}^{T} v_t = 0$ implies that the total amount is equivalent to $\sum_{t=1}^{T} \sum_{j=1}^{m_t} w_{cj}^t r_{ij}^t$ although the individual value of each site is concealed by v_{tc}. In this scheme, no site can reveal not only each cooccurrence information r_{ij}^t but also such other site-wise features as the actual value of $\sum_{j=1}^{m_t} w_{cj}^t r_{ij}^t$ and the number of items m_t on other sites. This encryption operation is performed for $i = 1, \ldots, n$, where different random vectors v_t should be used in each time so that site T cannot decode the vectors by cumulative calculation.

On the other hand, if we can share object memberships u_{ci}, item memberships w_{cj}^t can be calculated only with intra-site information in each site as follows:

$$w_{cj}^t = \frac{\exp\left(\dfrac{1}{\lambda_w} \sum_{i=1}^{n} u_{ci} r_{ij}^t\right)}{\displaystyle\sum_{\ell=1}^{m^t} \exp\left(\dfrac{1}{\lambda_w} \sum_{i=1}^{n} u_{ci} r_{i\ell}^t\right)}. \tag{5.9}$$

Here, it is also possible to use different λ_w for fuzzification degree in each site rather than collaborative use of a common degree in all sites. It is because the piece-wise constraint of $\sum_{j=1}^{m_t} w_{cj}^t = 1$ is independently forced to item memberships in each site while we just consider $\sum_{j=1}^{m} w_{cj} = 1$ in the whole data case.

Following the above consideration, FCCM for vertically distributed cooccurrence matrices is described as follows [7]:

[FCCM for Vertically Distributed Cooccurrence Matrices: FCCM-VD]

1. Given $n \times m_1$ matrix $R_1, \ldots, n \times m_T$ matrix R_T, and let C be the number of clusters. Choose the fuzzification weights λ_u and λ_w. (We can also use different λ_w in different sites because item memberships are available only in each site.)
2. **[Initialization]** Randomly initialize u_{ci} such that $\sum_{c=1}^{C} u_{ci} = 1$ and broadcast them to all sites.
3. **[Iterative process]** Iterate the following process until convergence of all u_{ci}.

   a. In sites $1, \ldots, T$, update w_{cj}^t using the current values of u_{ci}.
   b. For $i = 1, \ldots, n$
      i. In site 1, generate random vectors $v_t = (v_{t1}, \ldots, v_{tC})^\top$, $t = 1, \ldots, T$ such that $\sum_{t=1}^{T} v_t = 0$, and send v_t to site t.
      ii. In sites $1, \ldots, T$, calculate $v_{tc} + \sum_{j=1}^{m_t} w_{cj}^t r_{ij}^t$ and send them to site T.
      iii. In site T, update u_{ci} and broadcast them to all sites.
   c. Check the termination condition.

5.4 Collaborative Framework for FCCM of Horizontally Partitioned Cooccurrence Information

The secure process presented in the previous section can be extended to the case of handling horizontally distributed cooccurrence matrices. Assume that T sites $(t = 1, \ldots, T)$ share common m items $(j = 1, \ldots, m)$ and have different cooccurrence information on different objects, which are summarized into $n_t \times m$ matrices $R_t = \{r_{ij}^t\}$, where n_t is the number of objects in site t and $\sum_{t=1}^{T} n_t = n$. In this case, all sites are expected to share item memberships of co-clusters without revealing cooccurrence information and cluster memberships of individual objects, i.e., each site cannot broadcast each element of R_t and u_{ci}^t without encryption.

Then, the objective function to be maximized is modified as

$$J_{fccm-hd} = \sum_{c=1}^{C} \sum_{t=1}^{T} \sum_{i=1}^{n_t} \sum_{j=1}^{m} u_{ci}^t w_{cj} r_{ij}^t - \lambda_u \sum_{c=1}^{C} \sum_{t=1}^{T} \sum_{i=1}^{n_t} u_{ci}^t \log u_{ci}^t$$

$$- \lambda_w \sum_{c=1}^{C} \sum_{j=1}^{m} w_{cj} \log w_{cj}, \tag{5.10}$$

where u_{ci}^t is the object membership on object i in site t. Both sum-to-one conditions of FCCM can be utilized without modification as $\sum_{c=1}^{C} u_{ci}^t = 1, \forall i, t$, and $\sum_{j=1}^{m} w_{cj} = 1$, $\forall c$. Then, we can derive the equivalent result to the original FCCM algorithm under this situation.

Considering the optimality under the sum-to-one condition, the updating formula of object membership u_{ci}^t in site t is given as:

$$u_{ci}^t = \frac{\exp\left(\frac{1}{\lambda_u} \sum_{j=1}^{m} w_{cj} r_{ij}^t\right)}{\sum_{\ell=1}^{C} \exp\left(\frac{1}{\lambda_u} \sum_{j=1}^{m} w_{\ell j} r_{ij}^t\right)}. \tag{5.11}$$

If we can share the item memberships w_{cj} in all sites, object memberships u_{ci}^t can be calculated only with intra-site information and can keep secret in each site. Here, it is also possible to use different λ_u for fuzzification degree in each site rather than collaborative use of a common degree in all sites.

On the other hand, the updating formula of item memberships w_{cj} is given as:

$$w_{cj} = \frac{\exp\left(\frac{1}{\lambda_w} \sum_{t=1}^{T} \sum_{i=1}^{n_t} u_{ci}^t r_{ij}^t\right)}{\sum_{\ell=1}^{m^t} \exp\left(\frac{1}{\lambda_w} \sum_{t=1}^{T} \sum_{i=1}^{n_t} u_{ci}^t r_{i\ell}^t\right)}, \tag{5.12}$$

and must be calculated under collaboration of all sites.

In a similar manner to the vertically distributed data case, $\sum_{t=1}^{T} \sum_{i=1}^{n_t} u_{ci}^t r_{ij}^t$ can be securely calculated by adopting encryption operation without revealing the actual values of $\sum_{i=1}^{n_t} u_{ci}^t r_{ij}^t$. Assume that we have at least three sites, i.e., $T > 2$, and two sites of site 1 and site T are selected as representative sites.

- Site 1 generates a length m random vector $v_t = (v_{t1}, \ldots, v_{tm})^\top$ for each site t, such that $\sum_{t=1}^{T} v_t = 0$.
- Sites $1 \ldots T - 1$ send $v_{tj} + \sum_{i=1}^{n_t} u_{ci}^t r_{ij}^t$ to site T.
- Their total amount $\sum_{t=1}^{T} (v_{tj} + \sum_{i=1}^{n_t} u_{ci}^t r_{ij}^t)$ is calculated for estimating w_{cj} in site T.

$\sum_{t=1}^{T} v_t = 0$ implies that the total amount is equivalent to $\sum_{t=1}^{T} \sum_{i=1}^{n_t} u_{ci}^t r_{ij}^t$ although the individual value of each site is concealed by v_{tc}. In this scheme, no site can reveal the actual value of $\sum_{i=1}^{n_t} u_{ci}^t r_{ij}^t$ on other sites. This encryption operation is performed for $c = 1, \ldots, C$, where different random vectors v_t should be used in each time so that site T cannot decode the vectors by cumulative calculation.

Following the above consideration, FCCM for horizontally distributed cooccurrence matrices is described as follows:

[FCCM for Horizontally Distributed Cooccurrence Matrices: FCCM-HD]

1. Given $n_1 \times m$ matrix $R_1, \ldots, n_T \times m$ matrix R_T, and let C be the number of clusters. Choose the fuzzification weights λ_u and λ_w. (We can also use different λ_u in different sites because object memberships are available only in each site.)
2. **[Initialization]** Randomly initialize w_{cj} such that $\sum_{j=1}^{m} w_{cj} = 1$ and broadcast them to all sites.
3. **[Iterative process]** Iterate the following process until convergence of all w_{cj}.

   a. In sites $1, \ldots, T$, update u_{ci}^t using the current values of w_{cj}.
   b. For $c = 1, \ldots, C$
      i. In site 1, generate random vectors $v_t = (v_{t1}, \ldots, v_{tm})^\top$, $t = 1, \ldots, T$ such that $\sum_{t=1}^{T} v_t = 0$, and send v_t to site t.
      ii. In sites $1, \ldots, T$, calculate $v_{tj} + \sum_{i=1}^{n_t} u_{ci}^t r_{ij}^t$ and send them to site T.
      iii. In site T, update w_{cj} and broadcast them to all sites.
   c. Check the termination condition.

Additionally, handling horizontally distributed cooccurrence information can be utilized with large-scale databases as in the case of k-Means-type clustering if we can perform site-wise calculation in parallel.

5.5 Examples of FCCM Implementation with Distributed Cooccurrence Information

In order to demonstrate the advantages of collaborative analysis of distributed databases, an illustrative example is shown by using the cooccurrence information matrix used in the previous chapter. Figure 5.3a shows a cooccurrence information matrix $R = \{r_{ij}\}$ among 100 objects and 70 items, where black and white cells mean $r_{ij} = 1$ and $r_{ij} = 0$, respectively. In this data set, we can find roughly four co-clusters in the diagonal positions, which are somewhat sparse and are varied in noise.

Now, assume that the data matrix is vertically distributed in three sites as shown in Fig. 5.3b, where all sites have cooccurrence information on same 100 objects but independent m_t items such that $(m_1, m_2, m_3) = (25, 25, 20)$. In this experiment, the result of FCCM-VD is compared with those of the original (non-secure) FCCM and the site-wise independent analysis (without collaboration). The three types of FCCM implementations were performed with $\lambda_u = 0.1$ and $\lambda_w = 1.0$.

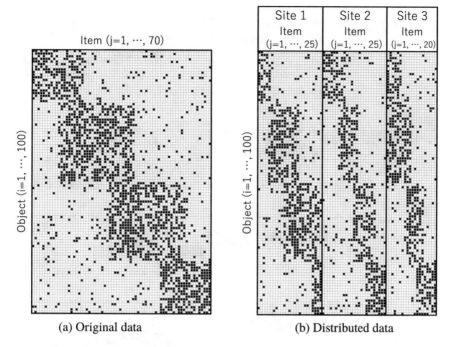

(a) Original data   (b) Distributed data

Fig. 5.3 Original and distributed cooccurrence information matrices

The original FCCM was implemented utilizing all elements of the original data without privacy preservation, whose results are shown in Figs. 5.4a and 5.5a. On the other hand, in the case of distributed databases, if we do not consider collaboration among sites, each site must perform FCCM clustering independently utilizing only the intra-site information. The site-wise results are shown in Figs. 5.4b and 5.5b, and, for comparison, the item memberships are also arranged to the original order in Fig. 5.5b'. The inferior results of each site imply that site-wise implementation with smaller databases degrades the performances of co-cluster structure analysis.

Then, the collaborative co-clustering was implemented with the FCCM-VD algorithm, whose results are shown in Figs. 5.4c and 5.5c, and, for comparison, the item memberships are also arranged to the original order in Fig. 5.5c'. Supported by collaborative implementation, we can derive almost equivalent results to the original non-secure ones. In this way, collaborative analysis is a powerful tool for improving data mining ability.

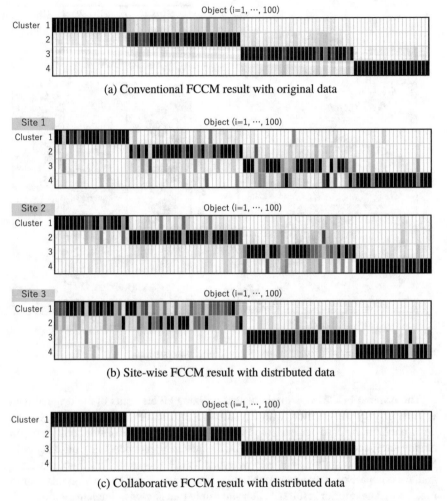

(a) Conventional FCCM result with original data

(b) Site-wise FCCM result with distributed data

(c) Collaborative FCCM result with distributed data

Fig. 5.4 Comparison of object memberships given with distributed cooccurrence information matrices

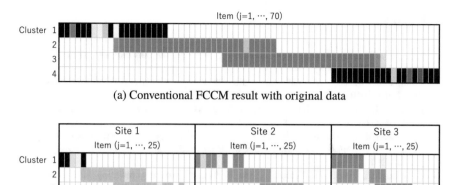

(a) Conventional FCCM result with original data

(b) Site-wise FCCM result with distributed data

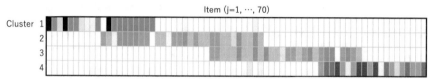

(b') Site-wise FCCM result with distributed data (arranged for comparison)

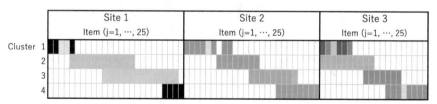

(c) Collaborative FCCM result with distributed data

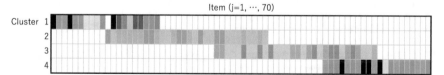

(c') Collaborative FCCM result with distributed data (arranged for comparison)

Fig. 5.5 Comparison of item memberships given with distributed cooccurrence information matrices

References

1. C.C. Aggarwal, P.S. Yu, *Privacy-Preserving Data Mining: Models and Algorithms* (Springer-Verlag, New York, 2008)
2. P. Samarati, Protecting respondents' identities in microdata release. IEEE Trans. Knowl. Data Eng. **13**(6), 1010–1027 (2001)
3. L. Sweeney, k-anonymity: a model for protecting privacy. Int. J. Uncertain., Fuzziness Knowl.-Based Syst. **10**(5), 557–570 (2002)
4. J.B. MacQueen, Some methods of classification and analysis of multivariate observations, in*Proceeding of 5th Berkeley Symposium on Mathematical Statistics and Probability* (1967), pp. 281–297
5. T.-K. Yu, D.T. Lee, S.-M. Chang, J. Zhan, Multi-party k-means clustering with privacy consideration, *Proceeding of the International Symposium on Parallel and Distributed Processing with Applications* (2010), pp. 200–207
6. F. Meskine, S.N. Bahloul, Privacy preserving k-means clustering: a survey research. Int. Arab. J. Inf. Technol. **9**(2), 194–200 (2012)
7. K. Honda, T. Oda, D. Tanaka, A. Notsu, A collaborative framework for privacy preserving fuzzy co-clustering of vertically distributed cooccurrence matrices, Adv. Fuzzy Syst., **2015**, #729072, 1–8 (2015)
8. J. Vaidya, C. Clifton, Privacy-preserving K-means clustering over vertically partitioned data, *Proceeding of the 9th ACM SIGKDD International Conference on Knowledge Discovery and Data Mining* (Washington, DC, USA 2003), pp. 206–215
9. S. Samet, A. Miri, L. Orozco-Barbosa, Privacy preserving k-means clustering in multi-party environment, in *Proceeding of the International Conference on Security and Cryptography* (2007), pp. 381–385
10. S. Jha, L. Kruger, P. Mcdaniel, Privacy preserving clustering, in *Proceeding of the 10th European Symposium On Research In Computer Security* (2005) pp. 397–417
11. C. Clifton, M. Kantarcioglu, J. Vaidya, X. Lin, M.Y. Zhu, Tools for privacy preserving distributed data mining. ACM SIGKDD Explor. Newsl. **4**(2), 28–34 (2002)
12. J. Vaidya, *A survey of privacy-preserving methods across vertically partitioned data, Privacy-preserving Data Mining: Models and Algorithms* (Springer, 2008), pp. 337–358
13. A. İnan, S.V. Kaya, Y. Saygın, E. Savaş, A.A. Hintoğlu, A. Levi, Privacy preserving clustering on horizontally partitioned data. Data & Knowl. Eng. **63**, 646–666 (2007)
14. J.C. Bezdek, *Pattern Recognition with Fuzzy Objective Function Algorithms* (Plenum Press, 1981)
15. T.C. Havens, J.C. Bezdek, C. Leckie, L.O. Hall, M. Palaniswami, Fuzzy c-means algorithms for very large data. IEEE Trans. Fuzzy Syst. **20**(6), 1130–1146 (2012)
16. N. Pal, J.C. Bezdek, Complexity reduction for large image processing, IEEE Trans. Syst., Man, Cybern., **32**(5), 598–611 (2002)
17. P. Hore, L. Hall, D. Goldgof, Single pass fuzzy c-means, in *Proceeding of IEEE International Conference on Fuzzy Systems* (2007), pp. 1–7
18. P. Hore, L. Hall, D. Goldgof, Y. Gu, A. Maudsley, A scalable framework for segmenting magnetic resonance images. J. Signal Process. Syst. **54**(1–3), 183–203 (2009)
19. C.-H. Oh, K. Honda, H. Ichihashi, Fuzzy clustering for categorical multivariate data, in *Proceeding of Joint 9th IFSA World Congress and 20th NAFIPS International Conference* (2001), pp. 2154–2159
20. S. Miyamoto, M. Mukaidono, Fuzzy c-means as a regularization and maximum entropy approach, in *Proceeding of the 7th International Fuzzy Systems Association World Congress*, vol. 2 (1997), pp. 86–92
21. S. Miyamoto, H. Ichihashi, K. Honda, *Algorithms for Fuzzy Clustering* (Springer, 2008)

Chapter 6
Three-Mode Fuzzy Co-clustering and Collaborative Framework

6.1 Introduction

As shown in the previous chapters, fuzzy co-clustering [1] is a basic but powerful tool for analyzing the intrinsic co-cluster structures varied in cooccurrence information among objects and items. However, if the cooccurrence features among objects and items are severely influenced by other intrinsic features, we should perform a collaborative analysis considering the relations among not only the two elements of objects and items but also other influential elements.

Let us consider an example of food preference analysis. We often want to reveal users' preferences on foods considering user-food cooccurrences. However, users' preferences can be influenced by implicit relation among users and cooking ingredients of each food even though ingredient information is not explicitly presented. Then, when we have not only cooccurrence information among users and foods but also intrinsic relation among foods and cooking ingredients, we can expect to find more useful food preference tendencies from three-mode cooccurrence information data.

In this chapter, a model of collaboratively utilizing three-mode cooccurrence information is introduced in the FCM-type co-clustering context, where FCM-like alternative optimization schemes [2] are performed considering cooccurrence relation among objects, items, and other ingredients. The FCCM algorithm [1] is extended to the three-mode FCCM (3FCCM) algorithm [3] by utilizing three types of fuzzy memberships for objects, items, and ingredients, where the aggregation degree of three features in each co-cluster is maximized through iterative updating of memberships.

Besides the development of various clustering algorithms, the awareness of information protection has been increasing in recent years [4] and privacy-preserving frameworks are necessary in performing clustering of real-world, large-scale data, which include personal information [5–10]. This chapter also considers a collaborative fuzzy co-clustering framework of three-mode cooccurrence information data as an extension of three-mode FCCM [11]. It is assumed that two types of cooccurrence information data of *objects* × *items* and *items* × *ingredients* are independently col-

© The Author(s), under exclusive license to Springer Nature Switzerland AG 2020
T.-C. T. Chen and K. Honda, *Fuzzy Collaborative Forecasting and Clustering*,
SpringerBriefs in Applied Sciences and Technology,
https://doi.org/10.1007/978-3-030-22574-2_6

lected and accumulated in different organizations, but the elements of cooccurrence information data cannot be disclosed each other from the viewpoint of information protection. The goal is to estimate co-cluster structures under collaboration of organizations such that we can extract the same co-cluster structures with the conventional three-mode FCCM keeping privacy preservation.

6.2 Extension of FCM-Type Co-clustering to Three-Mode Cooccurrence Data Analysis

Assume that we have cooccurrence information of n objects with m items and is summarized in an $n \times m$ matrix $R = \{r_{ij}\}$, where r_{ij} represents the cooccurrence degree among object i and item j. The goal of FCM-type fuzzy co-clustering is to find the dual partition of objects and items by estimating two types of fuzzy memberships [2]: u_{ci} for the membership of object i to cluster c and w_{cj} for the membership of item j to cluster c.

Fuzzy clustering for categorical multivariate data (FCCM) [1] defined the objective function considering the degree of aggregation of objects and items in each cluster and the entropy-based fuzzification [12, 13] as:

$$J_{fccm} = \sum_{c=1}^{C}\sum_{i=1}^{n}\sum_{j=1}^{m} u_{ci}w_{cj}r_{ij} - \lambda_u \sum_{c=1}^{C}\sum_{i=1}^{n} u_{ci}\log u_{ci} - \lambda_w \sum_{c=1}^{C}\sum_{j=1}^{m} w_{cj}\log w_{cj}. \quad (6.1)$$

Under the alternative optimization scheme like fuzzy c-Means (FCM) [2], u_{ci} and w_{cj} are iteratively updated until convergence.

Now, let us consider the situation where cooccurrence information among m items and p other ingredients is also available such that $m \times p$ matrix $S = \{s_{jk}\}$ is composed of the cooccurrence degree s_{jk} of item j and ingredient k and the ingredients can have some intrinsic influences on object–item cooccurrences. For example, in food preference analysis, R can be an evaluation matrix by n users (objects) on m foods (items) and S may be appearance/absence of p cooking ingredients (ingredients) in m foods (items). The goal of three-mode co-cluster analysis is to reveal the co-cluster structures among the objects, items, and ingredients considering R and S so that intrinsic relation among objects and ingredients can be utilized.

Assume that we have two cooccurrence information matrices shown in Table 6.1, where 1 implies such information as preference of food j by customer i or appearance of ingredient k in food j, respectively. If we arrange the matrices as Table 6.2, where we can find two object groups of {a, b, e} and {c, d, f} such that objects {a, b, e} are connected through items {1, 5, 6} while objects {c, d, f} are connected through items {2, 3, 4}. Additionally, ingredients {i, iii, v} and {ii, iv, vi} are also related to the two groups, respectively. This type of three-mode co-cluster structure implies the intrinsic *object–ingredient* connections such as {a, b, e} − {i, iii, v} and {c, d, f} − {ii, iv, vi}.

Table 6.1 A sample of three-mode cooccurrence information matrices

(a) object-item cooccurrences

Item		1	2	3	4	5	6
Object	a	1	0	0	0	0	1
	b	1	0	0	0	1	0
	c	0	1	1	0	0	0
	d	0	1	1	1	0	0
	e	0	0	0	0	1	1
	f	0	1	0	1	0	0

(b) item-ingredient cooccurrences

Ingredient		i	ii	iii	iv	v	vi
Item	1	1	0	1	0	0	0
	2	0	1	0	1	0	1
	3	0	0	0	1	0	1
	4	0	1	0	1	0	0
	5	1	0	1	0	1	0
	6	0	0	1	0	1	0

Table 6.2 Arranged matrices of Table 6.1

(a) Object-item cooccurrences

Item		1	5	6	2	3	4
Object	a	1	0	1	0	0	0
	b	1	1	0	0	0	0
	e	0	1	1	0	0	0
	c	0	0	0	1	1	0
	d	0	0	0	1	1	1
	f	0	0	0	1	0	1

(b) Item-ingredient cooccurrences

Ingredient		i	iii	v	ii	iv	vi
Item	1	1	1	0	0	0	0
	5	1	1	1	0	0	0
	6	0	1	1	0	0	0
	2	0	0	0	1	1	1
	3	0	0	0	0	1	1
	4	0	0	0	1	1	0

In order to extend the conventional FCCM algorithm to three-mode co-cluster analysis, additional memberships z_{ck} are introduced for representing the membership degree of ingredients k to co-cluster c. Besides the familiar pairs of objects and items simultaneously occur in a same cluster, typical ingredients of the representative items are expected to belong to the same cluster. Then, the aggregation degree to be maximized in the three-mode co-clustering [3] can be as:

$$Aggregation_c = \sum_{i=1}^{n} \sum_{j=1}^{m} \sum_{k=1}^{p} u_{ci} w_{cj} z_{ck} r_{ij} s_{jk}, \tag{6.2}$$

where each cluster should be composed of the group of familiar objects, items, and ingredients such that they are assigned to a same cluster when object i cooccurs with item j composed of ingredient k by implying an intrinsic connection between object i and ingredient k.

Here, we can adopt two different types of constraints to ingredient memberships z_{ck}, such that object-type probabilistic constraint $\sum_{c=1}^{C} z_{ck} = 1, \forall k$ or item-type typicality constraint $\sum_{k=1}^{p} z_{ck} = 1, \forall c$. In such cases as food preference analysis, some common ingredients may be widely used in many foods while other rare ingredients can be negligible in all clusters. Then, the memberships of ingredients should be considered not with intercluster assignment but with intra-cluster typicality. From the view point of typical ingredient selection for characterizing co-cluster features, item-type typicality constraint was adopted in [3], such that $\sum_{k=1}^{p} z_{ck} = 1, \forall c$.

6.2.1 Three-Mode Extension of FCCM

The FCCM algorithm was extended by using the modified aggregation criterion of Eq. (6.2) supported by the entropy-based fuzzification scheme. The objective function for three-mode FCCM (3FCCM) [3] is constructed by modifying the FCCM objective function as:

$$\begin{aligned}
J_{3fccm} = &\sum_{c=1}^{C} \sum_{i=1}^{n} \sum_{j=1}^{m} \sum_{k=1}^{p} u_{ci} w_{cj} z_{ck} r_{ij} s_{jk} \\
&- \lambda_u \sum_{c=1}^{C} \sum_{i=1}^{n} u_{ci} \log u_{ci} \\
&- \lambda_w \sum_{c=1}^{C} \sum_{j=1}^{m} w_{cj} \log w_{cj} \\
&- \lambda_z \sum_{c=1}^{C} \sum_{k=1}^{p} z_{ck} \log z_{ck},
\end{aligned} \tag{6.3}$$

where λ_z is the additional penalty weight for fuzzification of ingredient memberships z_{ck}. The larger the value of λ_z is, the fuzzier the ingredient memberships are.

The clustering algorithm is an iterative process of updating u_{ci}, w_{cj} and z_{ck} under the alternative optimization principle. Considering the necessary conditions for the optimality $\partial J_{3fccm}/\partial u_{ci} = 0$, $\partial J_{3fccm}/\partial w_{cj} = 0$ and $\partial J_{3fccm}/\partial z_{ck} = 0$ under the sum-to-one constraints, the updating rules for three types of memberships are given as:

$$u_{ci} = \frac{\exp\left(\frac{1}{\lambda_u} \sum_{j=1}^{m} \sum_{k=1}^{p} w_{cj} z_{ck} r_{ij} s_{jk}\right)}{\sum_{\ell=1}^{C} \exp\left(\frac{1}{\lambda_u} \sum_{j=1}^{m} \sum_{k=1}^{p} w_{\ell j} z_{\ell k} r_{ij} s_{jk}\right)}, \tag{6.4}$$

$$w_{cj} = \frac{\exp\left(\frac{1}{\lambda_w} \sum_{i=1}^{n} \sum_{k=1}^{p} u_{ci} z_{ck} r_{ij} s_{jk}\right)}{\sum_{\ell=1}^{m} \exp\left(\frac{1}{\lambda_w} \sum_{i=1}^{n} \sum_{k=1}^{p} u_{ci} z_{ck} r_{i\ell} s_{\ell k}\right)}, \tag{6.5}$$

and

$$z_{ck} = \frac{\exp\left(\frac{1}{\lambda_z} \sum_{i=1}^{n} \sum_{j=1}^{m} u_{ci} w_{cj} r_{ij} s_{jk}\right)}{\sum_{\ell=1}^{p} \exp\left(\frac{1}{\lambda_z} \sum_{i=1}^{n} \sum_{j=1}^{m} u_{ci} w_{cj} r_{ij} s_{j\ell}\right)}. \tag{6.6}$$

Following the above derivation, a sample algorithm is represented as follows.

[FCM-type Fuzzy Co-clustering for Three-mode Cooccurrence Data: 3FCCM]

1. Given $n \times m$ cooccurrence matrix R and $m \times p$ cooccurrence matrix S, and let C be the number of clusters. Choose the fuzzification weights λ_u, λ_w, and λ_z.
2. **[Initialization]** Randomly initialize u_{ci}, w_{cj}, and z_{ck}, such that $\sum_{c=1}^{C} u_{ci} = 1$, $\sum_{j=1}^{m} w_{cj} = 1$, and $\sum_{k=1}^{p} z_{ck} = 1$.
3. **[Iterative process]** Iterate the following process until convergence of all u_{ci}.

   a. Update u_{ci} with Eq. (6.4).
   b. Update w_{cj} with Eq. (6.5).
   c. Update z_{ck} with Eq. (6.6).
   d. Check the termination condition.

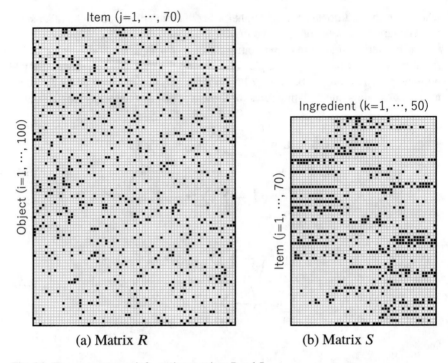

(a) Matrix R (b) Matrix S

Fig. 6.1 Two cooccurrence information matrices R and S

6.2.2 Examples of 3FCCM Implementation with Three-Mode Cooccurrence Information Data

In order to demonstrate the advantages of three-mode fuzzy co-clustering, an illustrative example is shown by using a toy example of three-mode cooccurrence information data. Assume that we have a cooccurrence information $R = \{r_{ij}\}$ among 100 objects and 70 items as shown in Fig. 6.1a, e.g., each of 100 customers chose some foods from 70 food menus. Black and white cells mean $r_{ij} = 1$ and $r_{ij} = 0$, respectively. Additionally, we also have mutual connections $S = \{s_{jk}\}$ among 70 items and 50 ingredients as shown in Fig. 6.1b, e.g., each of 70 foods contains some cooking ingredients. Black and white cells again mean $s_{jk} = 1$ and $s_{jk} = 0$, respectively.

At a glance, it is quite hard to find some co-cluster structures among objects, items, and ingredients from two cooccurrence information matrices R and S. In such tasks as food preference analysis, however, we can expect that each customer often choose some typical foods, which contains their favorite ingredients. In this experiment, an intrinsic preference $X = \{x_{ik}\}$ among 100 objects and 50 ingredients was assumed as in Fig. 6.2a, e.g., in purchase history transaction, $x_{ik} = 1$ if customer i prefers ingredient k. The three-mode cooccurrence matrices R and S of Fig. 6.1 were constructed such that $r_{ij} = 1$ when item j contains several favorite ingredients k such

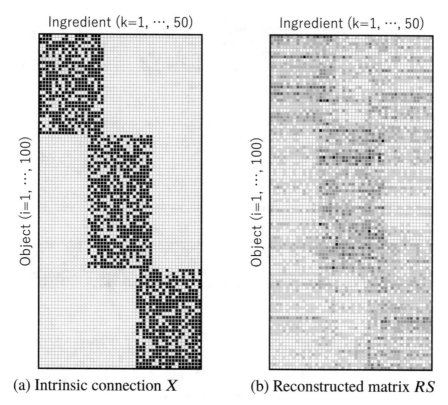

(a) Intrinsic connection X (b) Reconstructed matrix RS

Fig. 6.2 Intrinsic connection X and reconstructed matrix RS

as $x_{ik} = 1$. This intrinsic feature among objects and ingredients can be intuitively confirmed by multiplying two matrices $R \times S$ as shown in Fig. 6.2b, where the three co-clusters are weakly implied in diagonal locations like Fig. 6.2a. The goal of this experiment is to reveal the intrinsic three-mode co-cluster structure among objects, items, and ingredients from R and S.

6.2.2.1 Conventional FCCM Implementation

First, for comparison, the conventional FCCM was separately implemented with matrix R and matrix S using $\lambda_u = 0.1$ and $\lambda_w = 1.0$. The fuzzy memberships were estimated as shown in Figs. 6.3 and 6.4, respectively, where grayscale color depicts the degree of fuzzy memberships such as black for the maximum value and white for zero memberships. Because the conventional FCCM cannot consider the intrinsic connection among objects and ingredients, we cannot find co-cluster structures from both matrix R and matrix S.

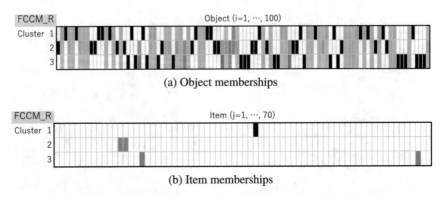

(a) Object memberships

(b) Item memberships

Fig. 6.3   FCCM memberships given from (object × item) matrix R

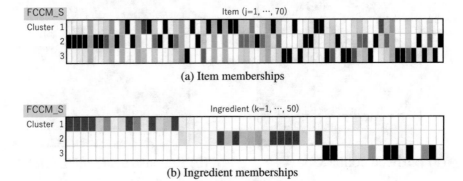

(a) Item memberships

(b) Ingredient memberships

Fig. 6.4   FCCM memberships given from (item × ingredient) matrix S

Besides, it is also unfavorable to implement the conventional FCCM to the reconstructed matrix $R \times S$. Figure 6.5 shows the clustering results given with $\lambda_u = 0.1$ and $\lambda_w = 10.0$. Due to heavy noise influences, the two types of memberships cannot work well for revealing the intrinsic co-cluster structures of the ideal matrix X.

The above results indicate that we need a novel method for analyzing three-mode cooccurrence information matrices.

6.2.2.2   3FCCM Implementation

Next, the 3FCCM algorithm was jointly implemented with the three-mode cooccurrence information matrices R and S. Because 3FCCM simultaneously considers the cooccurrences among objects, items, and ingredients, three types of fuzzy memberships are estimated at a time. Figure 6.6 shows the clustering results given with $\lambda_u = 0.05$, $\lambda_w = 0.1$, and $\lambda_z = 1.0$. Supported by the three-mode features, the 3FCCM algorithm could successfully extracted three object clusters, which were

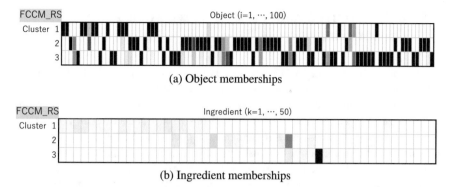

Fig. 6.5 FCCM memberships given from the reconstructed (object × ingredient) matrix RS

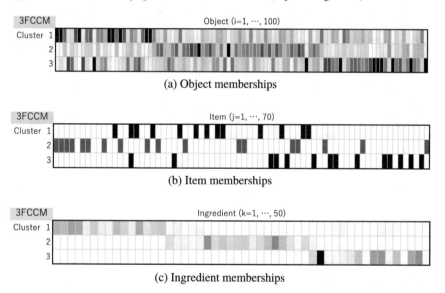

Fig. 6.6 3FCCM memberships given from three-mode cooccurrence information matrices R and S

implied in the intrinsic connection X, in conjunction with their typical ingredients. This successful result was achieved by focusing on some typical items in each cluster, which are emphasized with large memberships in Fig. 6.6b. For example, in food recommendation application, these typical foods can be selected for constructing an attractive menu considering intrinsic customer–ingredient connections.

6.3   Collaborative Framework for Three-Mode Fuzzy Co-clustering

Next, a novel framework for privacy preservation in three-mode FCCM is considered, where co-cluster estimation is jointly performed by two organizations [11]. Assume that organization A has $n \times m$ cooccurrence information $R = \{r_{ij}\}$ among n objects and m items, and organization B has $m \times p$ cooccurrence information $S = \{s_{jk}\}$ among m items and p ingredients, respectively. Although we can expect to estimate more informative co-cluster structures by adopting three-mode FCCM rather than the independent two-mode FCCM analysis in each organization, it may not be possible to disclose each element of cooccurrence matrices from the viewpoint of information protection.

For example, in food preference analysis, a sales outlet would store the preference relation among users and food menus while a caterer may have secret recipes on cooking ingredients of each food. Now, we can expect that preference tendencies among users and ingredients are useful for both the outlet and caterer in improving the quality of food recommendation and menu construction. However, it may be difficult for the two organizations to share their cooccurrence information matrices due to privacy or business issues. Collaborative framework for achieving three-mode co-clustering without data sharing is expected to bring a new business chance for both organizations under privacy preservation.

6.3.1   Collaborative Three-Mode FCCM

In the following, collaborative three-mode FCCM is considered [11], where the two organizations have the common goal of extracting co-cluster structures without disclosing each elements of cooccurrence matrices.

Now, organization A has *object* × *item* information while organization B has *item* × *ingredient* information. So, object memberships u_{ci} should be kept secret only in organization A, while ingredient memberships z_{ck} should be concealed only in organization B. By the way, item memberships w_{cj} can be shared by both organizations because the goal is to reveal the common item cluster structure.

First, in u_{ci} calculation, s_{jk} and z_{ck} are not directly available for organization A, where Eq. (6.4) is rewritten as:

$$u_{ci} = \frac{\exp\left(\frac{1}{\lambda_u} \sum_{j=1}^{m} w_{cj} r_{ij} \left(\sum_{k=1}^{p} z_{ck} s_{jk}\right)\right)}{\sum_{\ell=1}^{C} \exp\left(\frac{1}{\lambda_u} \sum_{j=1}^{m} w_{\ell j} r_{ij} \left(\sum_{k=1}^{p} z_{\ell k} s_{jk}\right)\right)}$$

$$= \frac{\exp\left(\frac{1}{\lambda_u} \sum_{j=1}^{m} w_{cj} r_{ij} \beta_{cj}\right)}{\sum_{\ell=1}^{C} \exp\left(\frac{1}{\lambda_u} \sum_{j=1}^{m} w_{\ell j} r_{ij} \beta_{\ell j}\right)}. \tag{6.7}$$

Next, in z_{ck} calculation, r_{ij} and u_{ci} are not directly available for organization B, where Eq. (6.6) is rewritten as:

$$z_{ck} = \frac{\exp\left(\frac{1}{\lambda_z} \sum_{j=1}^{m} w_{cj} s_{jk} \left(\sum_{i=1}^{n} u_{ci} r_{ij}\right)\right)}{\sum_{\ell=1}^{p} \exp\left(\frac{1}{\lambda_z} \sum_{j=1}^{m} w_{cj} s_{j\ell} \left(\sum_{i=1}^{n} u_{ci} r_{ij}\right)\right)}$$

$$= \frac{\exp\left(\frac{1}{\lambda_z} \sum_{j=1}^{m} w_{cj} s_{jk} \alpha_{cj}\right)}{\sum_{\ell=1}^{p} \exp\left(\frac{1}{\lambda_z} \sum_{j=1}^{m} w_{cj} s_{j\ell} \alpha_{cj}\right)}. \tag{6.8}$$

α_{cj} and β_{cj} are the following values calculated in organization A and organization B, respectively, which are referred to as shared information matrices A and B.

$$A = \begin{pmatrix} \alpha_{11} & \cdots & \alpha_{1m} \\ \vdots & \ddots & \vdots \\ \alpha_{C1} & \cdots & \alpha_{Cm} \end{pmatrix} = \begin{pmatrix} \sum_{i=1}^{n} u_{1i} r_{i1} & \cdots & \sum_{i=1}^{n} u_{1i} r_{im} \\ \vdots & \ddots & \vdots \\ \sum_{i=1}^{n} u_{Ci} r_{i1} & \cdots & \sum_{i=1}^{n} u_{Ci} r_{im} \end{pmatrix} \tag{6.9}$$

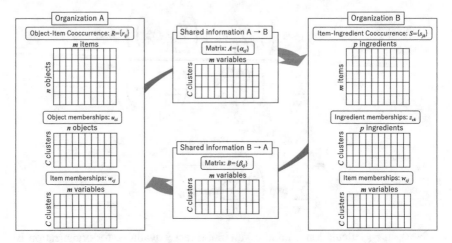

Fig. 6.7 An image of collaborative 3FCCM implementation

$$
B = \begin{pmatrix} \beta_{11} & \cdots & \beta_{1m} \\ \vdots & \ddots & \vdots \\ \beta_{C1} & \cdots & \beta_{Cm} \end{pmatrix} = \begin{pmatrix} \sum_{k=1}^{p} z_{1k}s_{1k} & \cdots & \sum_{k=1}^{p} z_{1k}s_{mk} \\ \vdots & \ddots & \vdots \\ \sum_{k=1}^{p} z_{Ck}s_{1k} & \cdots & \sum_{k=1}^{p} z_{Ck}s_{mk} \end{pmatrix} \tag{6.10}
$$

Finally, in w_{cj} calculation, Eq. (6.5) is rewritten as:

$$
\begin{aligned}
w_{cj} &= \frac{\exp\left(\dfrac{1}{\lambda_w}\left(\sum_{i=1}^{n} u_{ci}r_{ij}\right)\left(\sum_{k=1}^{p} z_{ck}s_{jk}\right)\right)}{\sum_{\ell=1}^{m}\exp\left(\dfrac{1}{\lambda_w}\left(\sum_{i=1}^{n} u_{ci}r_{i\ell}\right)\left(\sum_{k=1}^{p} z_{ck}s_{\ell k}\right)\right)} \\
&= \frac{\exp\left(\dfrac{1}{\lambda_w}\alpha_{cj}\beta_{cj}\right)}{\sum_{\ell=1}^{m}\exp\left(\dfrac{1}{\lambda_w}\alpha_{c\ell}\beta_{c\ell}\right)}.
\end{aligned} \tag{6.11}
$$

For example, an image of implementing collaborative 3FCCM can be visualized as in Fig. 6.7.

The above modification implies that it is possible to calculate fuzzy memberships in each organization by merely sharing the shared information matrices A and B without disclosing components of cooccurrence information matrices. The components of the shared information matrices A and B are regarded as the cluster structure information of items in organization A and organization B, respectively. α_{cj} represents the typicality of item j in cluster c taking object similarity into consideration while β_{cj} represents the typicality of item j in cluster c taking ingredient similarity into consideration. In these shared information, each of object and ingredient characteristics is kept secret because object similarity and ingredient similarity are shared after summing up in each organization. Actually, even if information matrix A is disclosed, organization B cannot know such knowledge as the number of objects n in organization A, and conversely, even if information matrix B is disclosed, organization A cannot know such knowledge as the number of ingredients p in organization B.

Following the above consideration, a sample algorithm is represented as follows.

[Collaborative FCM-type Fuzzy Co-clustering for Three-mode Cooccurrence Data: Collaborative 3FCCM]

1. Given $n \times m$ cooccurrence matrix R in organization A and $m \times p$ cooccurrence matrix S in organization B, and let C be the number of clusters. Choose the fuzzification weights λ_u, λ_w, and λ_z.
2. **[Initialization]** In organization B, randomly initialize w_{cj} such that $\sum_{j=1}^{m} w_{cj} = 1$, and calculate the components of shared information matrix B.
3. **[Iterative process]** Iterate the following process until convergence of all u_{ci}.

   a. In organization A, update u_{ci} with Eq. (6.7) and calculate information matrix A.
   b. In organization A, update w_{cj} with Eq. (6.11).
   c. From organization A to organization B, send information matrix A.
   d. In organization B, update z_{ck} with Eq. (6.8) and calculate information matrix B.
   e. In organization B, update w_{cj} with Eq. (6.11).
   f. From organization B to organization A, send information matrix B.
   g. In organization A, check the termination condition.

6.3.2 Examples of Implementation of Collaborative Framework for 3FCCM

Here, the previous example of Sect. 6.2.2 was again implemented with the collaborative framework for 3FCCM. It is noted that the collaborative framework is completely equivalent to the conventional 3FCCM algorithm without secure information sharing, i.e., information matrices A and B. Then, we can expect to derive a completely equivalent result through the two implementation strategies.

By the way, as an empirical study, let us consider the characteristic feature of the shared information matrices A and B, which represent the organization-wise item cluster structures estimated under collaboration. Figures 6.8 and 6.9 compare the shared information matrices A and B in the initial iteration and the final (20th in this experiment) iteration, respectively. Grayscale color depicts the element values such

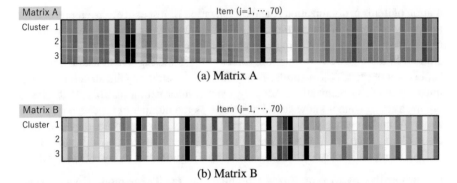

Fig. 6.8 Initial shared information matrices A and B in collaborative 3FCCM implementation

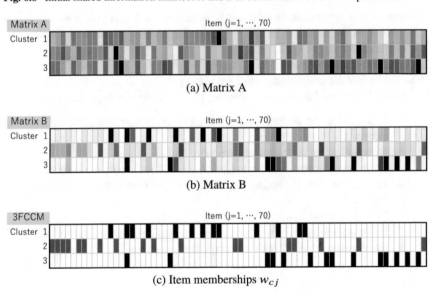

Fig. 6.9 Final shared information matrices A and B in collaborative 3FCCM implementation

as black for the maximum value and white for zero memberships. The initial structure of Fig. 6.8 forms almost random item selection in each organization, and matrices A and B have quite different features. On the other hand, after convergence, the organization-wise structures of Fig. 6.9a, b form still slightly different item clusters among two organizations such that some items have different typicalities. However, we should note that the final item memberships w_{cj} shown in Fig. 6.9c reflects the features of both organization-wise matrices A and B, which seem to be constructed through *max operation* among matrices A and B.

The collaborative process of homogenizing two matrices A and B is studied in Fig. 6.10, which depicts the trajectory of differences among two matrices until con-

Fig. 6.10 Trajectory of differences among two shared information matrices A and B in collaborative 3FCCM implementation

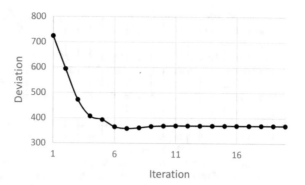

vergence. In order to compare two matrices, which are calculated under different scales, the sum of square deviations of elements was calculated after cluster-wise normalization so as to have zero means and unit variances:

$$Deviation = \sum_{c=1}^{C} \sum_{j=1}^{m} (\alpha_{cj} - \beta_{cj})^2. \tag{6.12}$$

Figure 6.10 implies that, starting from random assignment, the differences among two intra-structures A and B became smaller and are utilized for finding common co-cluster structures under collaboration. The collaborative 3FCCM algorithm gradually makes the two intra-structures homogeneous so that we can estimate three-mode co-cluster structures without disclosing original cooccurrence matrices R and S.

References

1. Oh, C.-H., Honda, K., Ichihashi, H, Fuzzy clustering for categorical multivariate data, Proc. of Joint 9th IFSA World Congress and 20th NAFIPS International Conference, 2154–2159 (2001)
2. J.C. Bezdek, *Pattern Recognition with Fuzzy Objective Function Algorithms* (Plenum Press, 1981)
3. Honda, K., Suzuki, Y., Ubukata, S., Notsu, A.: FCM-type fuzzy coclustering for three-mode cooccurrence data: 3FCCM and 3Fuzzy CoDoK, Advances in Fuzzy Systems, **2017**, #9842127, 1–8 (2017)
4. C.C. Aggarwal, P.S. Yu, *Privacy-Preserving Data Mining: Models and Algorithms* (Springer-Verlag, New York, 2008)
5. Vaidya, J., Clifton, C.: Privacy-preserving K-means clustering over vertically partitioned data, Proc. of the 9th ACM SIGKDD International Conference on Knowledge Discovery and Data Mining, 206–215, Washington, DC, USA (2003)
6. Jha, S., Kruger, L., Mcdaniel.: Privacy preserving clustering, Proc. of the 10th European Symposium On Research In Computer Security, 397–417 (2005)

7. Samet, S., Miri, A., Orozco-Barbosa, L.: Privacy preserving k-means clustering in multi-party environment. Proc. of the International Conference on Security and Cryptography, 381–385 (2007)
8. A. İnan, S.V. Kaya, Y. Saygın, E. Savaş, A.A. Hintoğlu, A. Levi, Privacy preserving clustering on horizontally partitioned data. Data & Knowledge Engineering **63**, 646–666 (2007)
9. Yu, T.-K., Lee, D.T., Chang, S.-M., Zhan, J.: Multi-party k-means clustering with privacy consideration, Proc. of the International Symposium on Parallel and Distributed Processing with Applications, 200–207 (2010)
10. Honda, K., Oda, T., Tanaka, D., Notsu, A.: A collaborative framework for privacy preserving fuzzy co-clustering of vertically distributed cooccurrence matrices, Advances in Fuzzy Systems, **2015**, #729072, 1–8 (2015)
11. Honda, K., Matsuzaki, S., Ubukata, S., Notsu, A.: Privacy preserving collaborative fuzzy co-clustering of three-mode cooccurrence data, Proc. of 15th International Conference on Modeling Decisions for Artificial Intelligence, **LNAI-11144**, 232–242, Springer (2018)
12. Miyamoto, S., Mukaidono, M.: Fuzzy c-means as a regularization and maximum entropy approach, Proc. of the 7th International Fuzzy Systems Association World Congress, **2**, 86–92 (1997)
13. S. Miyamoto, H. Ichihashi, K. Honda, *Algorithms for Fuzzy Clustering* (Springer, 2008)

Index

Printed in the United States
By Bookmasters